EUROPEAN SCIENCE
IN THE
SEVENTEENTH CENTURY

EUROPEAN SCIENCE
IN THE
SEVENTEENTH CENTURY

John Redwood

DAVID & CHARLES
Newton Abbot London Vancouver

BARNES & NOBLE BOOKS NEW YORK
(a division of Harper & Row Publishers, Inc)

ISBN 0 7153 7388 9 (Great Britain)
ISBN 0–06–495811–6 (United States)

This edition first published in 1977
in Great Britain by
David & Charles (Publishers) Limited
Brunel House Newton Abbot Devon

First published in the USA 1977 by
Harper & Row Publishers Inc
Barnes & Noble Import Division

Published in Canada by
Douglas David & Charles Limited
1875 Welch Street North Vancouver BC

Printed in Great Britain by
Latimer Trend & Company Ltd Plymouth

CONTENTS

INTRODUCTION

The seventeenth century was an age of intellectual revolt and reform as much as it was an age of political turmoil. It was an age which sought salvation through natural philosophy as well as through theology: men earnestly strove to understand the world about them by whatever means at their disposal. To many historians, the work of Pascal, Descartes, Mersenne, Harvey, Toricelli, Huyghens, Von Guericke, Halley, Newton and their colleagues amounted to a revolution of not inconsiderable proportions. Some, with Dijksterhuis, see seventeenth-century thinkers working towards the mechanical universe of Newton or Leibniz, undermining the Aristotelian universe based on forms and qualities and occult presences. The universe succumbed to mathematical treatment, and a world based on ideas and impressions was replaced by a world based upon matter, atoms and the propositions of dynamics. To other historians, less convinced of revolutions that take a century and movements that find unity in the most dissimilar of works and men, the seventeenth century remains an important focus of interest. For in the seventeenth century one does advance from Bacon, 'The bell that called the wits together', to Newton and his great scheme in the *Principia Mathematica*. It is true that it is by no means a straight advance from a world held together by the schoolman's logic to a world held together by the new Galilean dynamics, for the century of Galileo was also the century of Fludd, Van Helmont, and Paracelsus, who were not within the same rational mechanist tradition.

From the early days of Bacon and the patient English mathematicians of Gresham College in London to the flamboyance of the post-Restoration Royal Society, from the secret meetings of the early Italian groups under their Florentine patrons to the professional regulation of the Académie des Sciences, there was a stream of literature telling men how to think scientifically and recording the thoughts of those who had tried patent methods. It is this literature which concerns us in our study and from which this selection is drawn.

Our aim is to bring some of the many books, documents and papers relating to the natural philosophers and their world before the student and lay reader. The first section brings together some of the many writings of seventeenth-century men of letters and philosophy on the subject of aims and methods. It attempts to contrast the style of Francis Bacon and the Early English Empirical tradition with the more mathematical approach of Galileo, and with the rigours of the French Cartesian logic – a logic based upon Descartes's own belief in his triumph over scepticism in the *Discourse on Method*. It includes extracts from the works of Hobbes, whose severe mechanism and triumphant belief in the powers of matter and motion to explain the world led him into serious trouble with his contemporaries, and draws on the works of the later English tradition in the persons of Robert Boyle, Thomas Sprat, and Robert Hooke (who pioneered the use of the microscope). The section concludes with Newton's own views on reasoning and philosophy – the English reply to Cartesian domination of continental science.

The second section turns from these great arguments about the methods and competence of natural philosophy to look at the enterprises and achievements of the contemporary natural philosophers. In physics and mathematics, Napier is included for the great advance he made by developing the logarithm, a prerequisite for difficult calculations made later in the century. A preface to an edition of Euclid, the seminal mathematical text received from the Ancients by the seventeenth century, is also included, as is an example of the more vulgar interest in mathematics as a thing for toys and dilettante interest, seen here in Wilkins's *Mathematical Magick*. The section concludes with an example of the way in which mathematics and optics were strongly interrelated. Chemistry is represented, though sparsely, by Boyle's *The Sceptical Chymist* and magnetism by Gilbert's great work. The section on anatomy and medicine is a longer one, reflecting the overriding concern of seventeenth-century students of natural philosophy with the prolongation of life, with the alleviation of disease and with finding palliatives for pain and injury. Both anatomical and magical work are represented in the selection. The final subdivision of the second section is concerned with the great interest of seventeenth-century men in taxonomy and topography, represented by Ray, the great English naturalist, and Swammerdam, an important European student of biological life.

The third section attempts to display the range of institutional work in science and the journals which proliferated in the later seventeenth century, carrying the news of one group's activities to others and to gentlemen all across Western Europe wanting to be informed about the most modern developments in natural philosophy.

The final section was chosen in an attempt to show the importance of the microscope, developed in the seventeenth century, and to give brief biographical accounts of a few of the people most involved with the scientific movement.

This volume is compiled not in the belief that it is definitive or even comprehensive: merely in the belief that many of these texts and documents are among those vital to any understanding of seventeenth-century science, and that it is desirable that people interested in the subject should have ready access to them. These documents, I hope, display the diversity of intellectual endeavour and the range of intellectual achievement: the interrelationships of the different disciplines, and of the people and technology that made these studies possible. They attempt to display the diversity of natural philosophical studies in an age when the word science would be increasingly used but when its connotations would be much broader than those it carries in our own day. It attempts to demonstrate that while no specific discipline of the seventeenth-century mind was an isolated scientific enterprise, there was nevertheless something shared in common by all those who studied the countryside about them and the heavens above them, by all those who read Euclid, worshipped a Bacon, a Descartes or a Newton, and by those who believed in their euphoric prefaces that a new age of enlightenment was at hand.

Part One

AIMS AND METHODS

Throughout the seventeenth century scientists all over Europe were concerned about the nature and purpose of their studies. If they shared anything in common, it was their universal care to find the correct method of procedure: if their findings tended in any direction, it was towards the adoption of mathematical techniques in a wide range of works, and towards a more self-conscious experimental empiricism. There were dissenters, and there were many possible differences even within this broad framework of agreement; but the work of Huyghens and Newton, Descartes and Roberval, Pascal and Fermat, Toricelli and Mersenne, all agreed on the usefulness of mathematical models, while Hooke, Bacon, Sprat, Boyle, and others would stress their own brand of empirical enquiry.

All were united in attacking the work of their forbears: all railed against the claustrophobic air of the schoolroom, against the blinkered writings of academic translators and commentators annotating Aristotle or his minions. Bacon had pleaded for the amassing of evidence and for a new look at nature at the beginning of the seventeenth century, and in the 1620s in England men sought the libera philosophia. *They were unfair to Aristotle and the neo-Aristotelian tradition, unfair to the genuine achievement of the Merton school and the Paris school of the High Middle Ages, but they were sure that only through rejection and anger could they arrive at a new view of their universe. In the 1640s and 1650s it was the conjoint influences of Descartes and his method in France and Holland and the radical enlightenment of Comenius, Hartlib and Dury in revolutionary England that sought new paths. The continentals found in the logic of Descartes's* Principia *exciting new statements about the use of mathematics and the nature of philosophical logic: in England the educational reforms and the vulgar Baconian tradition brought back to England through the three European philosophers renewed English interest in natural philosophical problems. In the 1660s it was the turn of the scientific academies to demand a new freedom from prejudice and preconception. In the 1680s and 1690s it was the turn of Newton, Hooke and Huyghens to assert their independence in the realm of method, and lay down new standards and criteria for debate.*

The paradox of even this unity of outlook lay in the undoubted debt that the seventeenth-century men of science owed to their ancient predecessors, and even to Arab and medieval scholars. The later seventeenth century saw the great debate between the protagonists of the Ancients and the advocates of the Moderns. Men debated whether their own contemporaries had made a greater contribution to human knowledge and literature than had the giants of the Ancient heritage. Many thought that the Moderns were dwarfs standing on the shoulders of the giants, others felt that the Moderns had eclipsed their Ancient forbears, while others again felt that the Moderns were insignificant and puny beside the great work produced by the men of the Greek and Roman past. The paradox is that while the Moderns were defeating the Ancients in Will's Coffee House or in Paris, and while Wootten or Glanville, Fontenelle or their colleagues were attacking the last bastions of classical supremacy by arguments in favour of the Moderns, the classical heritage and classical models were settling more firmly on architecture, literature and science. The seventeenth century owed much to Aristotle, although it professed to attack him; it owed much to Pliny, to Pythagoras, to Euclid and to Plato. It exhumed the doctrines of the Platonists, which doctrines greatly influenced the Cambridge circle in England in the 1650s; it developed Aristotelian biology to greater heights in the Padua school; and it looked to Pliny for its information on natural history and for evidence for its rational theology. To Euclid and to Pythagoras the whole mathematical method owed more than a superficial resemblance.

While new sources among the Ancients were pillaged, or emphases changed in borrowing from the old, the age felt that it was free at last from centuries of darkness. It was reformation of the world of understanding: Bacon and his followers hauled down idols of the mind with as much willing iconoclasm as the Northern Church Fathers had mustered half a century or more before for the idols of the chapels and monasteries.

In their prefaces and prolegomena the men of science proclaimed old truths as new or relived old insights with new intensity. Great plans were afoot: Bacon and Hobbes felt no doubts about charting the whole of human learning, and Bacon, the Lord Chancellor, was sure that a few generations of hard work would bequeath to his successors the easy task of deducing the final axioms concerning the workings of the world. Descartes felt he had overcome the great sceptical crisis that threatened French if not European thought, and succeeded in finding a mechanical explanation of the world of matter which was deterministic and satisfyingly complete. Hobbes took mechanism further, and added to its bounds the activities of the mind: he felt no theological compunctions against explaining thought in terms of matter and motion, as well as so explaining the activities of the body and corporeal world. Robert Boyle summarised the excellence of this mechanical philosophy, which represented the triumph of self-conscious mathematical empiricism and the successful borrowing from a host of murky Classical authors in the Democritean and Epicurean traditions. Boyle himself, unlike Hobbes, was unwilling to explain everything in terms of matter and self-perpetuating motions: he agreed with the theologians that the universe demonstrated God's design, and believed that revelations and truths above reason should be respected and obeyed by mankind.

Pascal was not entirely convinced, either by the Cartesian dualism which argued that the world of matter was comprehensible to the rational intellect of man while the world of spirit was a thing apart, or by the English empirical and partly sceptical mixed tradition. Boyle shared some of Pascal's scruples, and to avoid the imputation of atomism, a physical doctrine tainted with atheism, renamed his theories the Corpuscularian Philosophy, thus steering clear of the unfortunate associations of Hobbes's doctrine of atomistic materialism. Newton proclaimed that he did not believe in hypotheses, and although few now believe him, he was careful to avoid publishing popular or general statements that could lead him into trouble with Church or Establishment. Newton believed in going only as far as the evidence would permit, and leaving the rest to the realm of theology or individual speculation.

Galileo had been responsible for much of the advance in the mathematical way of reasoning; contemporary with the Gresham group in England, and the first stirrings of the powerful French mathematical school under Mersenne, Galileo had successfully applied mathematics to astronomy and confirmed the revolution in the heavens. His views helped those sixteenth-century pioneers who had challenged the orthodoxy that the world was the centre of the universe and that the sun circulated around it. Under Galileo's impulse optics took a leap forward, and the telescope, at first a most inaccurate and unreliable instrument, was developed to help chart the newly enlarged heavens. Men became aware that the heavens were not circumscribed as the Ptolemaic universe of the medieval schoolroom suggested: that the universe did not end in the region of the fixed stars and the crystal sphere.

In the hands of Flamsteed or the Observatoire Royale observers telescopes became useful adjuncts to scientific enquiry, capable of a new degree of precision in measurement and in placing stars and other phenomena in the heavens. Similarly the world of the microscope attracted Leeuwenhoek and Power, Hooke and Malpighi, who used it to reveal a new world of complex smallness which captured the imagination of the Europeans. Leeuwenhoek working in Holland, backed by the Dutch glass industry and precision lens industry, Power and Hooke working in London in close conjunction with instrument makers, and Malpighi in Italy corresponding with the others across the Alps, made a new impact upon the world of the microcosm.

The patient empiricists of the Baconian or neo-Baconian tradition saw their rewards with the English Royal Society. Sprat, its apologist and historian, extols Francis Bacon, the great man who had first suggested an Academy of the Sciences. The first Curator of Experiments, Robert Hooke, provides a critique of this way of thinking, and rightly shows that practising scientists had come to realise the need for some speculation and mental organisation. He showed that Bacon's pure empiricism was not enough.

The debates over method were intense and never-ending. Everyone disagreed, and each piece of work was its own positive contribution to the debate over method and purpose. Early euphoria which saw unlimited potential in the power of science was transmuted to a more measured understanding of the limits as well as the possibilities of natural philosophy, though Newton believed in his God-given task of enlightening the world both with the true form of the heavens and with the arcana of nature's secrets. Some felt that in alchemy and occult arts the final power could be obtained; others preferred the success and glory of the mathe-matical method, which made such leaps and bounds in this century of discussion. Not all the prefaces and prologues should be taken at face value, for the works that followed often disappointed or differed from the programme outlined. But they should be read, for they capture the hope, the sense of vitality and freedom, the realisation of the new potential of understanding, the diversity of outlook concerning the way to enlighten-ment, such that perhaps no other succession of five generations has been able to equal.

FRANCIS BACON, 1561–1626

Bacon continued the attack of sixteenth-century reformers and new logi-cians, developing the criticisms of Ramus against what he took to be the errors of the Aristotelian school men. In his many books he suggested a new form of logic that should only make assertions about the natural world after compiling all the available information in large histories. He fought to eschew hypothesis and theory until men had seen nature anew and in the mass; he derided the old logicians for their syllogistic way of reasoning – based upon two premises with a conclusion following from them – and for conclusions ill grounded in the world of experience.

In the *proemium* to The Great Instauration (*1*) Bacon threw down the gauntlet to the old style of philosophising, accusing it of futility and the pursuit of chimaeras. These battle cries were to be reiterated time and again throughout the seventeenth century as men fought to free themselves from what they took to be the oppressive weight of medieval learning.

In the plan to the work Bacon explains his new method of logic (*2*). He was not suggesting an entirely mindless empiricism, as he demon-strates in the ensuing section in which he criticises the information found out by the senses. As the seventeenth century advanced and more reliance was placed on empirical study, the metaphysical and practical difficulties concerning sensory perception and the evidence obtained from the sense organs became more oppressive and more worrying.

Bacon gave a major impetus to natural philosophy. His great schemes were never fully realised, for it was a difficult task indeed to collate all the available evidence about the natural world in a set of encyclopaedic histories, and his method, which disliked theorising too soon, was im-

practicable. But his wish to find experiments that proved general axioms, and to pursue both fruit (tangible results) and light (knowledge), was influential both in England and in the rest of Europe through the mediation of Comenius, Dury and their school of reformers. It is to Bacon that we owe much of the interest in the history of trades and mechanical arts, the interest that prompted men to describe in detail the technology and method of working in a whole range of industries and skills. To Bacon the idea of the Academy was essential. Those early founders of the Royal Society may have been wrong to think that they were pursuing a true and pure Baconian path, but they found in Bacon a kindred spirit, whose ambitious plans represented a new programme for scientific enquiry. There were enough lines of questioning and sufficient unresolved problems in a work like Bacon's Sylva Sylvarum *to keep many a researcher in business for several decades. It was the fertility of Bacon's mind that furthered his influence and made him an attractive figure to the later natural philosophers.*

1 Francis Bacon
'THE GREAT INSTAURATION': PROEMIUM

Being convinced that the human intellect makes its own difficulties, not using the true helps which are at man's disposal soberly and judiciously; whence follows manifold ignorance of things, and by reason of that ignorance mischiefs innumerable; he thought all trial should be made, whether that commerce between the mind of man and the nature of things, which is more precious than anything on earth, or at least than anything that is of the earth, might by any means be restored to its perfect and original condition, or if that may not be, yet reduced to a better condition than that in which it now is. Now that the errors which have hitherto prevailed, and which will prevail for ever, should (if the mind be left to go its own way), either by the natural force of the understanding or by help of the aids and instruments of Logic, one by one correct themselves, was a thing not to be hoped for: because the primary notions of things which the mind readily and passively imbibes, stores up, and accumulates (and it is from them that all the rest flow) are false, confused and overhastily abstracted from the facts; nor are the secondary and subsequent notions less arbitrary and inconstant; whence it follows that the entire fabric of human reason which we employ in the inquisition of nature, is badly put together and built up, and like some magnificent structure without any foundation. For while men are occupied in admiring and applauding the false powers of the mind, they pass by and throw away those true powers, which, if it be supplied with the proper aids and can itself be content to wait upon nature instead of vainly affecting to overrule her, are within its reach. There was but one course left, therefore, – to try the whole thing anew upon a better plan, and to commence a total reconstruction of sciences, arts, and all human knowledge, raised upon the proper foundations. And this though in the project and undertaking it may seem a thing infinite and beyond the powers of man, yet when it comes to be dealt with it will be found sound and sober, more so than what has been done hitherto. For of this there is some issue; whereas in what is now done in the matter of science there is only a whirling round about, and perpetual agitation, ending where it began. And although he was well aware how solitary an enterprise it is, and how hard a thing to win faith and credit for, nevertheless he was resolved not to abandon either it or himself; nor to be deterred from trying and entering upon that one path which is alone open to the human mind. For better it is to make a beginning of that which may lead to something, than to engage in a perpetual struggle and pursuit in courses which have no exit. And certainly the two ways of contemplation are much like those two ways of action, so much celebrated in this – that the one, arduous and difficult in the beginning, leads out at last into the open country; while the other, seeming at first sight easy and free from obstruction, leads to pathless and precipitous places.

Moreover, because he knew not how long it might be before these things would occur to any one else, judging especially from this, that he has found no man hitherto who has applied his mind to the like, he resolved to publish at once so much as he has been able to complete. The cause of which haste was not ambition for himself, but solicitude for the work; that in case of his death there might remain some outline and project of that which he had conceived, and some evidence likewise of his honest mind and inclination towards the benefit of the human race. Certain it is that all other ambition whatsoever seemed poor in his eyes compared with the work which he had in hand; seeing that the matter at issue is either nothing, or a thing so great that it may well be content with its own merit, without seeking other recompence.

SOURCE: Spedding, Ellis and Heath (eds). *Francis Bacon: Works*, vol IV (London, 1860), 7–8

2 Francis Bacon
'THE GREAT INSTAURATION': PLAN

In accordance with this end is also the nature and order of the demonstrations. For in the ordinary logic almost all the work is spent about the syllogism. Of induction the logicians seem hardly to have taken any serious thought, but they pass it by with a slight notice, and hasten on to the formulæ of disputation. I on the contrary reject demonstration by syllogism, as acting too confusedly, and letting nature slip out of its hands. For although no one can doubt that things which agree in a middle term agree with one another (which is a proposition of mathematical certainty), yet it leaves an opening for deception; which is this. The syllogism consists of propositions; propositions of words; and words are the tokens and signs of notions. Now if the very notions of the mind (which are as the soul of words and the basis of the whole structure) be improperly and overhastily abstracted from facts, vague, not sufficiently definite, faulty in short in many ways, the whole edifice tumbles. I therefore reject the syllogism; and that not only as regards principles (for to principles the logicians themselves do not apply it) but also as regards middle propositions; which, though obtainable no doubt by the syllogism, are, when so obtained, barren of works, remote from practice, and altogether unavailable for the active department of the sciences. Although therefore I leave to the syllogism and these famous and boasted modes of demonstration their jurisdiction over popular arts and such as are matter of opinion (in which department I leave all as it is), yet in dealing with the nature of things I use induction throughout, and that in the minor propositions as well as the major. For I consider induction to be that form of demonstration which upholds the sense, and closes with nature, and comes to the very brink of operation, if it does not actually deal with it . . .

The sense fails in two ways. Sometimes it gives no information, sometimes it gives false information. For first, there are very many things which escape the sense, even when best disposed and no way obstructed; by reason either of the subtlety of the whole body, or the minuteness of the parts, or distance of place, or slowness or else swiftness of motion, or familiarity of the object, or other causes. And again when the sense does apprehend a thing its apprehension is not much to be relied upon. For the testimony and information of the sense has reference always to man, not to the universe; and it is a great error to assert that the sense is the measure of things.

To meet these difficulties, I have sought on all sides diligently and faithfully to provide helps for the sense – substitutes to supply its failures, rectifications to correct its errors; and this I endeavour to accomplish not so much by instruments as by experiments. For the subtlety of experiments is far greater than that of the sense itself, even when assisted by exquisite instruments; such experiments, I mean, as are skilfully and artificially devised for the express purpose of determining the point in question. To the immediate and proper perception of the sense therefore I do not give much weight; but I contrive that the office of the sense shall be only to judge of the experiment, and that the experiment itself shall judge of the thing.

SOURCE: as Document 1, 24–6

GALILEO GALILEI, 1564–1642

Galileo was a contemporary of Bacon's working in Italy during the early seventeenth century. Well known for his disputes with the Church, and for his enquiries with the newly invented telescope, Galileo was also of crucial importance in disseminating the idea of a mathematical base to natural speculation. Mathematical models were not new, for they owed something to the tradition of medieval kinematics passing through the Merton school of the thirteenth century to the Parisian and finally Italian schools of the Renaissance, just as they owed much to the influence of Galileo's father and to an early neo-platonic upbringing.

The method employed by Galileo has been much discussed, and the conclusion has emerged that Galileo was primarily an a priori logician and mathematician who sought crucial experiments by way of illustration of conclusions he had arrived at by other means. There is little truth in the famous story of the role of the Pisa tower experiment in disproving Aristotelian ideas on free fall, while attempts to make the necessary inclined planes for Galileo's brass ball experiments have been fraught with difficulties over friction and size. It seems that Galileo arrived at his conclusions more frequently by thought than by practice.

The publisher's introduction to the reader in the Dialogues concerning Two New Sciences *(3) represents a popular summary of his achievement as a mathematician and as an observer of the heavens. The introduction to the third day (4) represents a summary of Galileo's great achievement in bringing to life the new science of mechanics by giving a more systematic and informed treatment than his Mertonian or Parisian predecessors.*

Galileo should be remembered for his work in the Two New Sciences *perhaps more than for his* Starry Messenger *and his telescopic observations. But both sides of his personality illustrate well that conjoint influence of mathematics and empiricism on the seventeenth-century scientific mind that was to prove so successful in certain disciplines and fields of enquiry. In his* Il Saggiatore (Starry Messenger) *Galileo summarised an important message of the whole movement: that nature is mathematical, and therefore explicable by lemmae and propositions in a Euclidean fashion. It was this above all else that he attempted to do.*

3 Galileo
'THE TWO NEW SCIENCES': THE
PUBLISHER TO THE READER

Since society is held together by the mutual services which men render one to another, and since to this end the arts and sciences have largely contributed, investigations in these fields have always been held in great esteem and have been highly regarded by our wise forefathers. The larger the utility and excellence of the inventions, the greater has been the honor and praise bestowed upon the inventors. Indeed, men have even deified them and have united in the attempt to perpetuate the memory of their benefactors by the bestowal of this supreme honor.

Praise and admiration are likewise due to those clever intellects who, confining their attention to the known, have discovered and corrected fallacies and errors in many and many a proposition enunciated by men of distinction and accepted for ages as fact. Although these men have only pointed out falsehood and have not replaced it by truth, they are nevertheless worthy of commendation when we consider the well-known difficulty of discovering fact, a difficulty which led the prince of orators to exclaim: *Utinam tam facile possem vera reperire, quam falsa convincere.** And indeed, these latest centuries merit this praise because it is during them that the arts and sciences, discovered by the ancients, have been reduced to so great and constantly increasing perfection through the investigations and experiments of clear-seeing minds. This development is particularly evident in the case of the mathematical sciences. Here, without mentioning various men who have achieved success, we must without hesitation and with the unanimous approval of scholars assign the first place to Galileo Galilei, Member of the Academy of the Lincei. This he deserves not only because he has effectively demonstrated fallacies in many of our current conclusions, as is amply shown by his published works, but also

* Cicero. *de Natura Deorum*, I, 91. [*Trans.*]

because by means of the telescope (invented in this country but greatly perfected by him) he has discovered the four satellites of Jupiter, has shown us the true character of the Milky Way, and has made us acquainted with spots on the Sun, with the rough and cloudy portions of the lunar surface, with the three-fold nature of Saturn, with the phases of Venus and with the physical character of comets. These matters were entirely unknown to the ancient astronomers and philosophers; so that we may truly say that he has restored to the world the science of astronomy and has presented it in a new light.

Remembering that the wisdom and power and goodness of the Creator are nowhere exhibited so well as in the heavens and celestial bodies, we can easily recognize the great merit of him who has brought these bodies to our knowledge and has, in spite of their almost infinite distance, rendered them easily visible. For, according to the common saying, sight can teach more and with greater certainty in a single day than can precept even though repeated a thousand times; or, as another says, intuitive knowledge keeps pace with accurate definition.

But the divine and natural gifts of this man are shown to best advantage in the present work where he is seen to have discovered, though not without many labors and long vigils, two entirely new sciences and to have demonstrated them in a rigid, that is, geometric, manner: and what is even more remarkable in this work is the fact that one of the two sciences deals with a subject of never-ending interest, perhaps the most important in nature, one which has engaged the minds of all the great philosophers and one concerning which an extraordinary number of books have been written. I refer to motion [*moto locale*], a phenomenon exhibiting very many wonderful properties, none of which has hitherto been discovered or demonstrated by any one. The other science which he has also developed from its very foundations deals with the resistance which solid bodies offer to fracture by external forces [*per violenza*], a subject of great utility, especially in the sciences and mechanical arts, and one also abounding in properties and theorems not hitherto observed.

In this volume one finds the first treatment of these two sciences, full of propositions to which, as time goes on, able thinkers will add many more; also by means of a large number of clear demonstrations the author points the way to many other theorems as will be readily seen and understood by all intelligent readers.

SOURCE: Galileo Galilei. *A Discourse concerning Two New Sciences* (Ontario and London, 1954), xix–xxi

4 Galileo
'THE TWO NEW SCIENCES': THE THIRD DAY

CHANGE OF POSITION
[*De Motu Locali*]

My purpose is to set forth a very new science dealing with a very ancient subject. There is, in nature, perhaps nothing older than motion, concerning which the books written by philosophers are neither few nor small; nevertheless I have discovered by experiment some properties of it which are worth knowing and which have not hitherto been either observed or demonstrated. Some superficial observations have been made, as, for instance, that the free motion [*naturalem motum*] of a heavy falling body is continuously accelerated;* but to just what extent this acceleration occurs has not yet been announced; for so far as I know, no one has yet pointed out that the distances traversed, during equal intervals of time, by a body falling from rest, stand to one another in the same ratio as the odd numbers beginning with unity.

It has been observed that missiles and projectiles describe a curved path of some sort; however no one has pointed out the fact that this path is a parabola. But this and other facts, not few in number or less worth knowing, I have succeeded in proving; and what I consider more important, there have been opened up to this vast and most excellent science, of which my

* 'Natural motion' of the author has here been translated into 'free motion' – since this is the term used to-day to distinguish the 'natural' from the 'violent' motions of the Renaissance. [*Trans.*]

work is merely the beginning, ways and means by which other minds more acute than mine will explore its remote corners.

This discussion is divided into three parts; the first part deals with motion which is steady or uniform; the second treats of motion as we find it accelerated in nature; the third deals with the so-called violent motions and with projectiles.

UNIFORM MOTION

In dealing with steady or uniform motion, we need a single definition which I give as follows:

DEFINITION

By steady or uniform motion, I mean one in which the distances traversed by the moving particle during any equal intervals of time, are themselves equal.

CAUTION

We must add to the old definition (which defined steady motion simply as one in which equal distances are traversed in equal times) the word 'any,' meaning by this, all equal intervals of time; for it may happen that the moving body will traverse equal distances during some equal intervals of time and yet the distances traversed during some small portion of these time-intervals may not be equal, even though the time-intervals be equal.

From the above definition, four axioms follow [of which two are mentioned], namely:

AXIOM I

In the case of one and the same uniform motion, the distance traversed during a longer interval of time is greater than the distance traversed during a shorter interval of time.

AXIOM II

In the case of one and the same uniform motion, the time required to traverse a greater distance is longer than the time required for a less distance.

SOURCE: as Document 3, 153–4

26

RENÉ DESCARTES, 1596–1650

René Descartes was educated by French Jesuits, but later retreated to Holland to seek peace and philosophical quiet in Amsterdam. He perhaps was the most influential of all thinkers of the seventeenth century; his views were widely disseminated in France, the Low Countries and throughout the rest of Western Europe. His arguments became the bulwark of the continental school, based upon strong a priori *propositions arising out of a basic scepticism concerning the world.*

It was Descartes's Discourse on Method *which most captured the imagination of his contemporaries (5). Always reticent to publish anything, and worried about the reactions of the Catholic Church, Descartes nevertheless printed this part of a much larger work he had planned, to be called* De Mundo. *In the* Discourse *Descartes sets out arguments that he thinks satisfactorily settle the sceptical crisis of early seventeenth-century thought, finding certainty in the very fact that he is thinking about the problem at all. Descartes doubted the evidence of his senses, doubted his reason, doubted his own existence; but then discovered in his doubt proof that he must exist and that all else followed from that premise. His argument for the existence of God and for his own existence was accused of circularity by Arnauld and others, but Descartes was undeterred, for he had already discussed this accusation in his book and dismissed it. To him God's guarantee of reason and his own existence proved through doubting were precepts which reinforced one another: that they were mutually interdependent did not upset him. The views that he put forward were influential, and even in England, where a strong anti-Cartesian school developed, his arguments were frequently used to prove the existence of God.*

The scientific method Descartes advocated was mathematical. Like Galileo before him, he was fascinated by the potential of Euclid, geometry, and his own Cartesian algebra. He pioneered the Cartesian co-ordinates, representing algebraic equations by means of graphs, and transmuting the slopes of graphs back into algebraic equations. His general philosophical work was to become a standard classic of the scientific movement, as well as remaining one of the most interesting of autobiographies of a man who lived a life of the closet to the fullest and best advantage. The Discourse *is remarkable for the way in which it combines philosophical argument, belief in the faculties of man, and unerring testimony of Descartes's own personality, problems, and abilities.*

In the first part of the extract Descartes lays down the four guiding principles he used to further all his enquiries. They display his basic scepticism, and his fundamental faith in mathematics. The second part shows Descartes stumbling after truth, and the conclusion hints at the way in which he hopes his method will be useful in furthering the progress of the sciences.

5 René Descartes
'DISCOURSE ON METHOD'

When I was younger, I had studied a little logic in philosophy, and geometrical analysis and algebra in mathematics, three arts or sciences which would appear apt to contribute something towards my plan. But on examining them, I saw that, regarding logic, its syllogisms and most of its other precepts serve more to explain to others what one already knows, or even, like the art of Lully, to speak without judgement of those things one does not know, than to learn anything new. And although logic indeed contains many very true and sound precepts, there are, at the same time, so many others mixed up with them, which are either harmful or superfluous, that it is almost as difficult to separate them as to extract a Diana or a Minerva from a block of unprepared marble. Then, as for the geometrical analysis of the ancients and the algebra of the moderns, besides the fact that they extend only to very abstract matters which seem to be of no practical use, the former is always so tied to the inspection of figures that it cannot exercise the understanding without greatly tiring the imagination, while, in the latter, one is so subjected to certain rules and numbers that it has become a confused and obscure art which oppresses the mind instead of being a science which cultivates it. This was why I thought I must seek some other method which, while continuing the advantages of these three, was free from their defects. And as a multiplicity of laws often furnishes excuses for vice, so that a State is much better ordered when, having only very few laws, they are very strictly observed, so, instead of this great number of precepts of which logic is composed, I believed I would have sufficient in the four following rules, so long as I took a firm and constant resolve never once to fail to observe them.

The first was never to accept anything as true that I did not know to be evidently so: that is to say, carefully to avoid precipitancy and prejudice, and to include in my judgements nothing more than what presented itself so clearly and so distinctly to my mind that I might have no occasion to place it in doubt.

The second, to divide each of the difficulties that I was examining into as many parts as might be possible and necessary in order best to solve it.

The third, to conduct my thoughts in an orderly way, beginning with the simplest objects and the easiest to know, in order to climb gradually, as by degrees, as far as the knowledge of the most complex, and even supposing some order among those objects which do not precede each other naturally.

And the last, everywhere to make such complete enumerations and such general reviews that I would be sure to have omitted nothing.

These long chains of reasonings, quite simple and easy, which geometers are accustomed to using to teach their most difficult demonstrations, had given me cause to imagine that everything which can be encompassed by man's knowledge is linked in the same way, and that, provided only that one abstains from accepting any for true which is not true, and that one always keeps the right order for one thing, to be deduced from that which precedes it, there can be nothing so distant that one does not reach it eventually, or so hidden that one cannot discover it. And I was in no great difficulty in seeking which to begin with because I knew already that it was with the simplest and easiest to know; and considering that, among all those who have already sought truth in the sciences, only the mathematicians have been able to arrive at any proofs, that is to say, certain and evident reasons, I had no doubt that it was by the same things which they had examined that I should begin, although I did not expect any other usefulness from this than to accustom my mind to nourish itself on truths and not to be content with false reasons . . .

. . . I had long ago noticed that, in matters relating to conduct, one needs sometimes to follow, just as if they were absolutely indubitable, opinions one knows to be very unsure, as has been said above; but as I wanted to concentrate solely on the search for truth, I thought I ought to do just the opposite, and reject as being absolutely false everything in which I could suppose the slightest reason for doubt, in order to see if there did not remain after that anything in my belief which was entirely indubitable. So, because our senses sometimes play us false, I decided to suppose that there was nothing at all which was such as they cause us to imagine it; and because there are men who make mistakes in reasoning, even with the simplest geometrical matters, and make paralogisms, judging that I was as liable to error as anyone else, I rejected as being false all the reasonings I had hitherto accepted as proofs. And finally, considering that all the same thoughts that we have when we are awake can also come to us when we are asleep, without any one of them then being true, I resolved to pretend that nothing which had ever entered my mind was any more true than the illusions of my dreams. But immediately afterwards I became aware that, while I decided thus to think that everything was false, it followed necessarily that I who thought thus must be something; and observing that this truth: *I think, therefore I am*, was so certain and so evident that all the most extravagant suppositions of the sceptics were not capable of shaking it, I judged that I could accept it without scruple as the first principle of the philosophy I was seeking.

Then, examining attentively what I was, and seeing that I could pretend that I had no body and that there was no world or place that I was in, but that I could not, for all that, pretend that I did not exist, and that, on the contrary, from the very fact that I thought of doubting the truth of other things, it followed very evidently and very certainly that I existed; while, on the other hand, if I had only ceased to think, although all the rest of what I had ever imagined had been true, I would have had no reason to believe that I existed; I thereby concluded that I was a substance, of which the whole essence or nature consists in thinking, and which, in order to exist, needs no place and depends on no material thing; so that this 'I', that is to say, the mind, by which I am what I am, is entirely distinct from the body, and even that it is easier to know than the body, and moreover, that even if the body were not, it would not cease to be all that it is.

After this, I considered in general what is needed for a proposition to be true and certain; for, since I had just found one which I knew to be so, I thought that I ought also to know what this certainty consisted of. And having noticed that there is nothing at all in this, *I think, therefore I am*, which assures me that I am speaking the truth, except that I see very clearly that in order to think one must exist, I judged that I could take it to be a general rule that the things we conceive very clearly and very distinctly are all true, but that there is nevertheless some difficulty in being able to recognize for certain which are the things we see distinctly . . .

In conclusion, I do not wish to speak here in detail of the progress I hope to make in the sciences in the future nor to make any promise to the public that I am not certain of being able to fulfil; but I will say simply that I have resolved to devote the time left to me to live to no other occupation than that of trying to acquire some knowledge of Nature, which may be such as to enable us to deduce from it rules in medicine which are more assured than those we have had up to now; and that my inclination turns me away so strongly from all other sorts of projects, and particularly from those which can only be useful to some while being harmful to others, that if any situation arose in which I was forced to engage in such matters, I do not think that I would be able to succeed. On this, I here make a public declaration which I know very well cannot serve to make me of consequence in the world, but then I have no wish to be so; and I shall always hold myself more obliged to those by whose favour I enjoy my leisure unhindered, than to those who might offer me the highest dignities on earth.

SOURCE: René Descartes. *Discourse on Method* (London, 1970), 40–2, 49–50, 53–5, 91

THOMAS HOBBES, 1588–1679

Thomas Hobbes was primarily a philosopher, but no philosopher of the mid-seventeenth century could remain uninfluenced by natural philosophy, or by the political events of the civil wars in England and elsewhere. There were revolts not only of the English Parliament against the English crown but also the Frondes in France and revolts of the outlying provinces against the Spanish crown in Italy, Andalusia, Catalonia and Portugal. Hobbes knew Gassendi and his circle in France, a group of men putting forward an opposing atomistic doctrine to that advocated by Descartes. He met Galileo on a trip through France and Italy, for he frequently travelled as tutor to English gentlemen. He was in contact with the English group of scientists in Oxford and London during the civil wars.

Hobbes was a keen mathematician, engaging in an acrimonious dispute with John Wallis, the Oxford professor of geometry, and writing on a wide range of topics; he pursued an atomist's line of enquiry, after the fashion of Gassendi and the early English circle which had centred around Harriot and Northumberland in the early years of the seventeenth century. Hobbes was much criticised for his mathematics, attempting as he had to square the circle. Nevertheless mathematics was an important element in Hobbes's philosophy, even if his competence at mathematics was brought into serious question by the strictures of Wallis.

He was famous most of all for his great book entitled Leviathan. *This book put forward a mechanical simile for the functioning of the Christian commonwealth, comparing it with the mechanical workings of the human body. Hobbes extended mechanical and mathematical determinism to its limits, believing apparently that everything could be explained in terms of necessary motions of atoms within bodies. The introduction to* Leviathan *(6) elaborates the meaning of the title, and provides a vivid portrait of a mechanical system, a political framework conditioned by the new science.*

Hobbes's political theory was based upon an understanding of the functionings of the atoms making up the body, and the operation of forces conditioning the pattern of mental and physical phenomena. In the section on reason and science (7) Hobbes voices his scathing dislike of his medieval forbears, and escoriates most brutally the prejudices and errors of previous men of learning. He used this attack to illuminate the value of his own method of scientific enquiry all the more clearly.

The final extract from his work (8) shows one of the many seventeenth-century schemes illustrating the way in which the different branches of knowledge were thought to relate one to the other. It is interesting for the way in which it divides science into two realms: the first concerned with natural philosophical questions, a more usual definition of scientific endeavour, and the second concerned with politics and civil philosophy, a discipline which Hobbes himself attempted to make as rigorous and as mathematically based as other branches of knowledge had become. Hobbes thought he could build a science of human behaviour on the basis of advances made in the understanding of the physical world. His works from which these extracts are taken attempt to provide a coherent view of psychology, philosophy, political society, mathematics, and the natural world.

6 Thomas Hobbes
'LEVIATHAN': INTRODUCTION

NATURE, the art whereby God hath made and governs the world, is by the art of man, as in many other things, so in this also imitated, that it can make an artificial animal. For seeing life is but a motion of limbs, the beginning whereof is in some principal part within; why may we not say, that all automata (engines that move themselves by springs and wheels as doth a watch) have an artificial life? For what is the heart, but a spring; and the nerves, but so many strings; and the joints, but so many wheels, giving motion to the whole body, such as was intended by the artificer? Art goes yet further, imitating that rational and most excellent work of nature, man. For by art is created that great LEVIATHAN called a COMMONWEALTH, or STATE, in Latin CIVITAS, which is but an artificial man; though of greater stature and strength than the natural, for whose protection and defence it was intended; and in which the sovereignty is an artificial soul, as giving life and motion to the whole body; the magistrates, and other officers of judicature and execution, artificial joints; reward and punishment, by which fastened to the seat of the sovereignty every joint and member is moved to perform his duty, are the nerves, that do the same in the body natural; the wealth and riches of all the particular members, are the strength; *salus populi*, the people's safety, its business; counsellors, by whom all things needful for it to know are suggested unto it, are the memory; equity, and laws, an artificial reason and will; concord, health; sedition, sickness; and civil war, death. Lastly, the pacts and covenants, by which the parts of this body politic were at first made, set together, and united, resemble that fiat, or the *let us make man*, pronounced by God in the creation.

To describe the nature of this artificial man, I will consider.

First, the matter thereof, and the artificer; both which is man.

Secondly, how, and by what covenants it is made; what are the rights and just power or authority of a sovereign; and what it is that preserveth or dissolveth it.

Thirdly, what is a Christian commonwealth.

Lastly, what is the kingdom of darkness.

Concerning the first, there is a saying much usurped of late, that wisdom is acquired, not by reading of books, but of men. Consequently whereunto, those persons, that for the most part can give no other proof of being wise, take great delight to show what they think they have read in men, by uncharitable censures of one another behind their backs. But there is another saying not of late understood, by which they might learn truly to read one another, if they would take the pains; that is *nosce teipsum*, read thyself: which was not meant, as it is now used, to countenance, either the barbarous state of men in power, towards their inferiors; or to encourage men of low degree, to a saucy behaviour towards their betters; but to teach us, that for the similitude of the thoughts and passions of one man, to the thoughts and passions of another, whosoever looketh into himself, and considereth what he doth, when he does think, opine,

34

reason, hope, fear, &c. and upon what grounds; he shall thereby read and know, what are the thoughts and passions of all other men upon the like occasions. I say the similitude of passions, which are the same in all men, desire, fear, hope, &c.; not the similitude of the objects of the passions, which are the things desired, feared, hoped, &c.: for these the constitution individual, and particular education, do so vary, and they are so easy to be kept from our knowledge, that the characters of man's heart, blotted and confounded as they are with dissembling, lying, counterfeiting, and erroneous doctrines, are legible only to him that searcheth hearts. And though by men's actions we do discover their design sometimes; yet to do it without comparing them with our own, and distinguishing all circumstances, by which the case may come to be altered, is to decypher without a key, and be for the most part deceived, by too much trust, or by too much diffidence; as he that reads, is himself a good or evil man.

But let one man read another by his actions never so perfectly, it serves him only with his acquaintance, which are but few. He that is to govern a whole nation, must read in himself, not this or that particular man; but mankind: which though it be hard to do, harder than to learn any language or science; yet when I shall have set down my own reading orderly, and perspicuously, the pains left another, will be only to consider, if he also find not the same in himself. For this kind of doctrine admitteth no other demonstration.

SOURCE: Molesworth (ed). *Thomas Hobbes: Works*, vol III (London, 1839), ix–xii

7 Thomas Hobbes
'LEVIATHAN': OF REASON AND SCIENCE

WHEN a man reasoneth, he does nothing else but conceive a sum total, from addition of parcels; or conceive a remainder, from subtraction of one sum from another; which, if it be done by words, is conceiving of the consequence of the names of all the parts, to the name of the whole; or from the names of the whole and one part, to the name of the other part. And though in some things, as in number, besides adding and subtracting, men name other operations, as multiplying and dividing, yet they are the same; for multiplication, is but adding together of things equal; and division, but subtracting of one thing, as often as we can. These operations are not incident to numbers only, but to all manner of things that can be added together, and taken one out of another. For as arithmeticians teach to add and subtract in numbers; so the geometricians the same in lines, figures, solid and superficial, angles, proportions, times, degrees of swiftness, force, power, and the like; the logicians teach the same in consequences of words; adding together two names to make an affirmation, and two affirmations to make a syllogism; and many syllogisms to make a demonstration; and from the sum, or conclusion of a syllogism, they subtract one proposition to find the other. Writers of politics add together pactions to find men's duties; and lawyers, laws and facts, to find what is right and wrong in the actions of private men. In sum, in what matter soever there is place for addition and subtraction, there also is place for reason; and where these have no place, there reason has nothing at all to do.

Out of all which we may define, that is to say determine, what that is, which is meant by this word reason, when we reckon it amongst the faculties of the mind. For REASON, in this sense, is nothing but reckoning, that is adding and subtracting, of the consequences of general names agreed upon for the marking and signifying of our thoughts; I say marking them when we reckon by ourselves, and signifying, when we demonstrate or approve our reckonings to other men.

. . . For it is most true that Cicero saith of them somewhere; that there can be nothing so absurd, but may be found in the books of philosophers. And the reason is manifest. For there is not one of them that begins his ratiocination from the definitions, or explications of the names they are to use; which is a method that hath been used only in geometry; whose conclusions have thereby been made indisputable.

I. The first cause of absurd conclusions I ascribe to the want of method; in that they begin not their ratiocination from definitions; that is, from settled significations of their words; as if they could cast account, without knowing the value of the numeral words, one, two and three.

And whereas all bodies enter into account upon divers considerations, which I have mentioned in the precedent chapter; these considerations being diversely named, divers absurdities proceed from the confusion and unfit connexion of their names into assertions. And therefore,

II. The second cause of absurd assertions, I ascribe to the giving of names of bodies to accidents; or of accidents to bodies; as they do, that say, faith is infused, or inspired; when nothing can be poured, or breathed into anything, but body; and that, extension is body; that phantasms are spirits, &c.

III. The third I ascribe to the giving of the names of the accidents of bodies without us, to the accidents of our own bodies as they do that say, the colour is in the body; the sound is in the air, &c.

IV. The fourth, to the giving of the names of bodies to names, or speeches; as they do that say, that there be things universal; that a living creature is genus, or a general thing, &c.

V. The fifth, to the giving of the names of accidents to names and speeches; as they do that say, the nature of a thing is its definition; a man's command is his will; and the like.

VI. The sixth, to the use of metaphors, tropes, and other rhetorical figures, instead of words proper. For though it be lawful to say, for example, in common speech, the way goeth, or leadeth hither, or thither; the proverb says this or that, whereas ways cannot go, nor proverbs speak; yet in reckoning, and seeking of truth, such speeches are not to be admitted.

VII. The seventh, to names that signify nothing; but are taken up, and learned by rote from the schools, as hypostatical, transubstantiate, consubstantiate, eternal-now, and the like canting of schoolmen.

To him that can avoid these things it is not easy to fall into any absurdity, unless it be by the length of an account; wherein he may perhaps forget what went before. For all men by nature reason alike, and well, when they have good principles. For who is so stupid, as both to mistake in geometry, and also to persist in it, when another detects his error to him? . . .

To conclude, the light of human minds is perspicuous words, but by exact definitions first snuffed, and purged from ambiguity; reason is the pace; increase of science, the way; and the benefit of mankind, the end. And, on the contrary, metaphors, and senseless and ambiguous words, are like *ignes fatui*; and reasoning upon them is wandering amongst innumerable absurdities; and their end, contention and sedition, or contempt . . .

The signs of science are some, certain and infallible; some, uncertain. Certain, when he that pretendeth the science of any thing, can teach the same; that is to say, demonstrate the truth thereof perspicuously to another; uncertain, when only some particular events answer to his pretence, and upon many occasions prove so as he says they must. Signs of prudence are all uncertain; because to observe by experience, and remember all circumstances, that may alter the success, is impossible. But in any business, whereof a man has not infallible science to proceed by; to forsake his own natural judgment, and be guided by general sentences read in authors, and subject to many exceptions, is a sign of folly, and generally scorned by the name of pedantry. And even of those men themselves, that in councils of the commonwealth love to show their reading of politics and history, very few do it in their domestic affairs, where their particular interest is concerned; having prudence enough for their private affairs: but in public they study more the reputation of their own wit, than the success of another's business.

SOURCE: as Document 6, 29–38

8 Thomas Hobbes
'LEVIATHAN': OF THE SEVERAL SUBJECTS
OF KNOWLEDGE

THERE are of KNOWLEDGE two kinds; whereof one is knowledge of fact: the other knowledge of the consequence of one affirmation to another. The former is nothing else, but sense and memory, and is absolute knowledge; as when we see a fact doing, or remember it done: and this is the knowledge required in a witness. The latter is called science; and is conditional; as when we know, that, if the figure shown be a circle, then any straight line through the centre shall divide it into two equal parts. And this is the knowledge required in a philosopher; that is to say, of him that pretends to reasoning.

The register of knowledge of fact is called history. Whereof there be two sorts: one called natural history; which is the history of such facts, or effects of nature, as have no dependence on man's will; such as are the histories of metals, plants, animals, regions, and the like. The other, is civil history; which is the history of the voluntary actions of men in commonwealths.

The registers of science, are such books as contain the demonstrations of consequences of one affirmation, to another; and are commonly called books of philosophy; whereof the sorts are many, according to the diversity of the matter; and may be divided in such manner as I have divided them in the following table [see pp 40 and 41].

SOURCE: as Document 6, 71-3

Consequences from the accidents common to all bodies natural; which are *quantity*, and *motion*

Consequences from the accidents of bodies natural; which is called NATURAL PHILOSOPHY.

Consequences from the qualities of bodies *transient*, such as sometimes appear, sometimes vanish, *Meteorology*.

PHYSICS or consequences from *qualities*.

Consequences from the qualities of the *stars*.

Consequences from the qualities of bodies *permanent*.

Consequences of the qualities from *liquid* bodies, that fill the space between the stars; such as are the *air*, or substances ethereal.

SCIENCE, that is, knowledge of consequences; which is called also PHILOSOPHY.

Consequences from the qualities of *bodies terrestrial*.

Consequences from the accidents of *politic* bodies; which is called POLITICS, and CIVIL PHILOSOPHY.

1. Of consequences from the *institution* of COMMONWEALTH, to the *rights*, and *duties* of the *body politic* or *sovereign*.
2. Of consequences from the same, to the *duty* and *right* of the *subjects*.

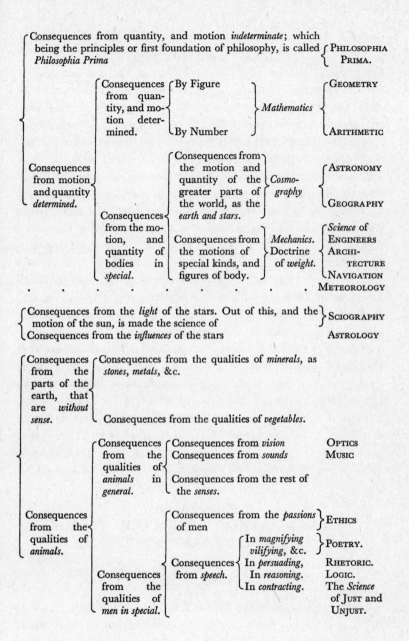

Consequences from quantity, and motion *indeterminate*; which being the principles or first foundation of philosophy, is called *Philosophia Prima* — PHILOSOPHIA PRIMA.

Consequences from motion and quantity *determined.*

Consequences from quantity, and motion determined. — By Figure / By Number — *Mathematics* — GEOMETRY / ARITHMETIC

Consequences from the motion, and quantity of bodies in *special.*

Consequences from the motion and quantity of the greater parts of the world, as the *earth and stars.* — *Cosmography* — ASTRONOMY / GEOGRAPHY

Consequences from the motions of special kinds, and figures of body. — *Mechanics.* Doctrine of *weight.* — *Science* of ENGINEERS ARCHITECTURE NAVIGATION

METEOROLOGY

Consequences from the *light* of the stars. Out of this, and the motion of the sun, is made the science of — SCIOGRAPHY

Consequences from the *influences* of the stars — ASTROLOGY

Consequences from the parts of the earth, that are *without sense.*

Consequences from the qualities of *minerals*, as *stones, metals*, &c.

Consequences from the qualities of *vegetables.*

Consequences from the qualities of *animals.*

Consequences from the qualities of *animals* in general.

Consequences from *vision* — OPTICS
Consequences from *sounds* — MUSIC
Consequences from the rest of the *senses.*

Consequences from the qualities of *men in special.*

Consequences from the *passions* of men — ETHICS

Consequences from *speech.* — In *magnifying* vilifying, &c. — POETRY.
In *persuading,* — RHETORIC.
In *reasoning.* — LOGIC.
In *contracting.* — The *Science* of JUST and UNJUST.

41

ROBERT HOOKE, 1635–1703

Robert Hooke was one of the first full-time salaried scientists with an institutional patron, holding the post of Curator of Experiments at the Royal Society; he had also worked with Robert Boyle. In his employment with the Royal Society it was Hooke's task to find things to amuse and inform the gentlemen each week at their meetings, and to construct crucial or diverting experiments that would be clearly visible in the lecture room.

In consequence Hooke developed many divergent natural philosophical interests. He was concerned with astronomy, the motion of the planets, the form of the heavens, and with problems concerned with matter, force, and the physical shape of the universe. But among all his interests his concern with microscopy became the one for which he is in some ways most famous. He contributed greatly to the development of the microscope and especially to the successful use of compound lens models. In the preface to his Micrographia *we see him delivering a kind of warning to the Royal Society, in which he suggests that self-conscious Baconian ideals are all very well but scarcely practicable (9). While some gentlemen seemed to think with Bacon that their task is to heap experiment on experiment, observation upon observation, until all is clear, Hooke thinks that there is a role for hypotheses in discovery. He finds it necessary in his studies to use themes and ideas to organise his material and findings. The argument is especially interesting coming from Hooke, for it serves as an admonition to the less professional of his fellow Society men that the Baconian method in its purest form would produce neither the light nor the fruit that they anticipated.*

9 Robert Hooke
'MICROGRAPHIA': DEDICATION TO THE ROYAL SOCIETY AND PREFACE

DEDICATION

After my Address to our Great Founder and Patron, I could not but think my self oblig'd, in consideration of those many Ingagements you have laid upon me, to offer these my poor Labours to this Most Illustrious Assembly. You have been pleas'd formerly to accept of these rude Draughts. I have since added to them some Descriptions, and some Conjectures of my own. And therefore, together with your Acceptance, I must also beg your pardon. The Rules you have prescrib'd your selves in your Philosophical Progress do seem the best that have ever yet been practis'd. And particularly that of avoiding Dogmatizing, and the espousal of any Hypothesis not sufficiently grounded and confirm'd by Experiments. This way seems the most excellent, and may preserve both Philosophy and Natural History from its former Corruptions. In saying which, I may seem to condemn my own Course in this Treatise; in which there may perhaps be some Expressions, which may seem more positive then your Prescriptions will permit: And though I desire to have them understood only as Conjectures and Queries (which your Method does not altogether disallow) yet if even in those I have exceeded, 'tis fit that I should declare, that it was not done by your Directions. For it is most unreasonable, that you should undergo the imputation of the faults of my Conjectures, feeling you can receive so small advantage of reputation by the sleight Observations of Robert Hooke.

PREFACE

It is the great prerogative of Mankind above other Creatures, that we are not only able to behold the works of Nature, or barely to sustein our lives by them, but we have also the power of considering, comparing, altering, assisting, and improving them to various uses. And as this is the peculiar privilege of humane Nature in general, so is it capable of being so far advanced by the helps of Art, and Experience, as to make some Men excel others in their Observations, and Deductions, almost as much as they do Beasts. By the addition of such artificial Instruments and methods, there may be, in some manner, a reparation made for the mischiefs, and imperfection, mankind has drawn upon it self, by negligence, and intemperance, and a wilful and superstitious deserting the Prescripts and Rules of Nature, whereby every man, both from a deriv'd corruption, innate and born with him, and from his breeding and converse with men, is very subject to slip into all sorts of error.

The only way which now remains for us to recover some degree of those former perfections, seems to be, by rectifying the operations of the Sense, the Memory, and Reason, since upon the evidence, the strength, the integrity, and the right correspondence of all these, all the light, by which our actions are to be guided, is to be renewed, and all our command over things is to be establisht.

It is therefore most worthy of our consideration, to recollect their several defects, that so we may the better understand how to supply them, and by what assistances we may inlarge their power, and secure them in performing their particular duties.

As for the actions of our Senses, we cannot but observe them to be in many particulars much outdone by those of other Creatures, and when at best, to be far short of the perfection they seem capable of: And these infirmities of the Senses arise from a double cause, either from the disproportion of the Object to the Organ, whereby an infinite number of things can never enter into them, or else from error in the Perception, that many things, which come within their reach, are not received in a right manner.

The like frailties are to be found in the Memory; we often let many things slip away from us, which deserve to be retain'd; and of those which we treasure up, a great part is either frivolous or false; and if good, and substantial, either in tract of time obliterated, or at best so overwhelmed and buried under more frothy notions, that when there is need of them, they are in vain fought for.

The two main foundations being so deceivable, it is no wonder, that all the succeeding works which we build upon them, of arguing, concluding, defining, judging, and all the other degrees of Reason, are lyable to the same imperfection, being, at best, either vain, or uncertain: So that the errors of the understanding are answerable to the two other, being defective both in the quantity and goodness of its knowledge; for the limits, to which our thoughts are confin'd, are small in respect

44

of the vast extent of Nature it self; some parts of it are too large to be comprehended, and some too little to be perceived. And from thence it must follow, that not having a full sensation of the object, we must be very lame and imperfect in our conceptions about it, and in all the propositions which we build upon it; hence we often take the shadow of things for the substance, small appearances for good similitudes, similitudes for definitions; and even many of those, which we think to be the most solid definitions, are rather expressions of our own misguided apprehensions then of the true nature of the things themselves.

The effects of these imperfections are manifested in different ways, according to the temper and disposition of the several minds of men, some they incline to gross ignorance and stupidity, and others to a presumptuous imposing on other mens Opinions, and a confident dogmatizing on matters, whereof there is no assurance to be given.

Thus all the uncertainty, and mistakes of humane actions, proceed either from the narrowness and wandring of our Senses, from the slipperiness or delusion of our Memory, from the confinement or rashness of our Understanding, so that 'tis no wonder, that our power over natural causes and effects is so slowly improv'd, seeing we are not only to contend with the obscurity and difficulty of the things whereon we work and think, but even the forces of our own minds conspire to betray us.

These being the dangers in the process of humane Reason, the remedies of them all can only proceed from the real, the mechanical, the experimental Philosophy, which has this advantage over the Philosophy of discourse and disputation, that whereas that chiefly aims at the subtilty of its Deductions and Conclusions, without much regard to the first ground-work, which ought to be well laid on the Sense and Memory; so this intends the right ordering of them all, and the making them serviceable to each other.

The first thing to be undertaken in this weighty work, is a watchfulness over the failings and an inlargement of the dominion, of the Senses.

To which end it is requisite, first, That there should be a

scrupulous choice, and a strict examination, of the reality, constancy, and certainty of the Particulars that we admit: This is the first rise whereon truth is to begin, and here the most severe, and most impartial diligence, must be imployed; the storing up of all, without any regard to evidence or use, will only tend to darkness and confusion. We must not therefore esteem the riches of our Philosophical treasure by the number only, but chiefly by the weight; the most vulgar Instances are not to be neglected, but above all, the most instructive are to be entertain'd; the footsteps of Nature are to be trac'd, not only in her ordinary course, but when she seems to be put to her shifts, to make many doublings and turnings, and to use some kind of art in indeavouring to avoid our discovery.

The next care to be taken, in respect of the Senses, is a supplying of their infirmities with Instruments, and, as it were, the adding of artificial Organs to the natural; this in one of them has been of late years accomplisht with prodigious benefit to all sorts of useful knowledge, by the invention of Optical Glasses. By the means of Telescopes, there is nothing so far distant but may be represented to our view; and by the help of Microscopes, there is nothing so small, as to escape our inquiry; hence there is a new visible World discovered to the understanding. By this means the Heavens are open'd, and a vast number of new Stars, and new Motions, and new Productions appear in them, to which all the antient Astronomers were utterly Strangers. By this the Earth it self, which lyes so neer us, under our feet, shews quite a new thing to us, and in every particle of its matter, we now behold almost as great a variety of Creatures, as we were able before to reckon up in the whole Universe it self.

It seems not improbable, but that by these helps the subtilty of the composition of Bodies, the structure of their parts, the various texture of their matter, the instruments and manner of their inward motions, and all the other possible appearances of things, may come to be more fully discovered; all which the antient Peripateticks were content to comprehend in two general and (unless further explain'd) useless words of Matter and Form. From whence there may arise many admirable advantages, towards the increase of the Operative, and the Mechanick Knowledge, to which this Age seems so much inclined, because we may perhaps be inabled to discern all the secret workings of Nature, almost in the same manner as we do those that are the productions of Art, and are manag'd by Wheels, and Engines, and Springs, that were devised by humane Wit . . .

The truth is, the Science of Nature has been already too long made only a work of the Brain and the Fancy: It is now high time that it should return to the plainness and soundness of Observations on material and obvious things. It is said of great Empires, That the best way to preserve them from decay, is to bring them back to the first Principles, and Arts, on which they did begin. The same is undoubtedly true in Philosophy, that by wandring far away into invisible Notions, has almost quite destroy'd it self, and it can never be recovered, or continued, but by returning into the same sensible paths, in which it did at first proceed.

If therefore the Reader expects from me any infallible Deductions, or certainty of Axioms, I am to say for my self, that those stronger Works of Wit and Imagination are above my weak Abilities; or if they had not been so, I would not have made use of them in this present Subject before me: Wherever he finds that I have ventur'd at any small Conjectures, at the causes of the things that I have observed, I beseech him to look upon them only as doubtful Problems, and uncertain ghesses, and not as unquestionable Conclusions, or matters of uncon-futable Science; I have produced nothing here, with intent to bind his understanding to an implicit consent; I am so far from that, that I desire him, not absolutely to rely upon these Observations of my eyes, if he finds them contradicted by the future Ocular Experiments of sober and impartial Discoverers. . .

The last indeed is the most hazardous Enterprize, and yet the most necessary; and that is, to take such care that the Judgment and the Reason of Man (which is the third Faculty to be repair'd and improv'd) should receive such assistance, as to avoid the dangers to which it is by nature most subject. The Imperfections, which I have already mention'd, to which it is lyable, do either belong to the extent or the goodness of its knowledge; and here the difficulty is the greater, least that which may be thought a remedy for the one should prove destructive to the other, least by seeking to inlarge our Knowledge, we should render it weak and uncertain; and least by being too scrupulous and exact about every Circumstance of it, we should confine and streighten it too much.

What kind of mechanical way, and physical invention is there requir'd, that might not this way be found out? The Invention of a way to find the Longitude of places is easily perform'd, and that to as great perfection as is desir'd, or to as great an accurateness as the Latitude of places can be found at Sea; and perhaps yet also to a greater certainty then that has been hitherto found, as I shall very speedily freely manifest to the world. The way of flying in the Air seems principally unpracticable, by reason of the want of strength in humane muscles; if therefore that could be suppli'd, it were, I think, easie to make twenty contrivances to perform the office of Wings: What Attempts also I have made for the supplying that Defect, and my successes therein, which, I think, are wholly new, and not inconsiderable, I shall in another place relate.

'Tis not unlikely also, but that Chymists, if they followed this method, might find out their so much sought for Alkahest. What an universal Menstruum, which dissolves all sorts of Sulphureous Bodies, I have discover'd (which has not been before taken notice of as such) I have shewn in the sixteenth Observation.

What a prodigious variety of Inventions in Anatomy has this latter Age afforded, even in our own Bodies, in the very Heart, by which we live, and the Brain, which is the seat of our knowledge of other things? witness all the excellent Works of Pecquet, Bartholinus, Billius, and many others; and at home, of Doctor Harvy, Doctor Ent, Doctor Willis, Doctor Glisson. In Celestial Observations we have far exceeded all the Antients, even the Chaldeans and Egyptians themselves, whose vast

Plains, high Towers, and clear Air, did not give them so great advantages over us, as we have over them by our Glasses. By the help of which, they have been very much outdone by the famous Galileo, Hevelius, Zulichem; and our own Country-men, Mr. Rook, Doctor Wren, and the great Ornament of our Church and Nation, the Lord Bishop of Exeter. And to say no more in Aerial Discoveries, there has been a wonderful progress made by the Noble Engine of the most Illustrious Mr. Boyle, whom it becomes me to mention with all honour, not only as my particular Patron, but as the Patron of Philosophy it self; which he every day increases by his Labours, and adorns by his Example . . .

And if they will please to take any incouragement from so mean and so imperfect endeavours as mine, upon my own experience, I can assure them, without arrogance, That there has not been any inquiry or Problem in Mechanicks, that I have hitherto propounded to my self, but by a certain method (which I may on some other opportunity explain) I have been able presently to examine the possibility of it; and if so, as easily to excogitate divers wayes of performing it: And indeed it is possible to do as much by this method in Mechanicks, as by Algebra can be perform'd in Geometry. Nor can I at all doubt, but that the same method is as applicable to Physical Enquiries, and as likely to find and reap thence as plentiful a crop of Inventions; and indeed there seems to be no subject so barren, but may with this good husbandry be highly improv'd.

SOURCE: Robert Hooke. *Micrographia* (London, 1665)

ROBERT BOYLE, 1627–91

Robert Boyle was a wide-ranging scientist and philosopher whose interests lay in the reconciliation of science with religion, and in propa-gandising the usefulness and excellence of the new mechanical philosophy. Boyle, the son of a nobleman with some independent means, worked away in Oxford and then in London. He is perhaps best known for his experiments with the air pump and his work on pressure in gases, leading to his formulation of the law which bears his name. But Boyle's interests were protean, ranging from alchemy through the physics of gases and pressures to the importance of mechanical disciplines in natural philosophical studies. Boyle was a keen advocate of the history of trades project for the Royal Society, and was himself an active member of that club for the advancement of learning.

In the extract below (10) Boyle argues that mechanical disciplines are useful to natural philosophy itself, and urges a rapprochement between the two kinds of activity to their mutual benefit. Boyle acknowledges his debt both to Mersenne, the French scientific intermediary, correspondent, and mathematician in his own right, and to Archimedes, the ancient mathematician and physicist from whom the seventeenth century drew so many ideas. Boyle believes that the mariner's compass and the air gun are but two examples of the progress that can be recorded by following his advice concerning the rapprochement of philosophy and mechanics. He also foresees the productive interrelations between mechanics in its broadest sense and natural philosophy in its purest.

10 Robert Boyle
'OF THE USEFULNESS OF MECHANICAL DISCIPLINES TO NATURAL PHILOSOPHY'
SHEWING, THAT THE POWER OF MAN MAY BE PROMOTED BY THE NATURALIST'S SKILL IN MECHANICKS

TO prevent the danger of stumbling (as they speak) at the threshold, I shall begin this discourse with advertising you, that I do not here take the term mechanicks in that stricter and more proper sense, wherein it is wont to be taken, when it is used only to signify the doctrine about the moving powers (as the beam, the leaver, the screws, and the wedge,) and of framing engines to multiply force: but I here understand the word mechanicks in a larger sense, for those disciplines, that consist of the applications of pure mathematicks to produce or modify motion in inferior bodies: so that in this sense they comprise not only the vulgar staticks, but divers other disciplines, such as the centrobaricks, hydraulicks, pneumaticks, hydrostaticks, balisticks, &c. the etymology of whose names may inform you about what subjects they are conversant.

Now that these arts (if you will allow them that name) may be of great use to the experimental philosopher, and assist him to enlarge the empire of man, may be made probable by this general consideration, that divers of those things, which in the former essay have been evinced to make the mathematicks useful to the naturalist, may be applied *mutatis mutandis* to the mechanicks also. Besides, that these disciplines have some advantages peculiar to themselves. But the truth of what is thus represented in general terms will possibly be better discern'd, and more persuasive, if we descend to some particulars.

I. FIRST then, the phænomena afforded us by these arts ought to be looked upon as really belonging to the history of nature in its full and due extent. And therefore as they fall under the cognizance of the naturalist, and challenge his speculation; so it may well be supposed, that being throughly understood, they cannot but much contribute to the advancement of his knowledge, and consequently of his power, which we have often observed to be grounded upon his knowledge, and proportionate to it. When, for instance, we see a piece of wood, ducked under water, emerge again and float, even vulgar naturalists think, that it belongs to them to consider the reason of this emersion and floating, which they endeavour to render from the positive levity, which they fancy to be (upon the account of the air and fire) inherent in the wood, though some woods, that will swim in water, being put into oil, or high rectified spirit of wine, may sink.

But I see not, why it should not belong to philosophers to consider and investigate the reason, why one part of floating wood appears above the water, whilst the other keeps beneath it; and why the extent part is equal to the immersed, or either greater or lesser than it, in such a determinate proportion; and why the same wood will sink deeper in some waters than in others, (as in a river than in the sea) as on the other side some woods will sink lower than others in the same water. For if these things to be duly examined, as they may by the help of hydrostaticks, not only the cause of these and the like phænomena will be discovered; but by the applications of that discovery an easy way may be devised to measure and estime at the differing strength of several salt springs, and also of divers kinds of lixiviums, and brines; to which may be added divers other practical corollaries from the same discoveries, which I shall hereafter have occasion to particularize.

II. The mechanical disciplines help me to devise and judge of such hypotheses, as relate to those subjects, wherein the notions and theorems of mechanicks either ought necessarily to be considered, or may usefully be so.

Of this we have instances, not only in those engines that are artificial, and are looked upon as purely mechanical, as the screw, the crane, the ballance, &c. but in many familiar phænomena, in which the theorems of mechanicks are not wont to be taken notice of to have an interest. As in the carrying a pike or musket on one's shoulder, in the force of strokes with a longer or shorter sword or other instrument, the taking up and the holding a pike or sword at arms-length, and the power, that a rudder has to steer a ship; in rowing with boats, in breaking of sticks against one's knee, and in a multitude of other familiar instances, of which the naturalist's skill in mechanicks will enable him to give a far more clear and solid account, than the ancient schoolmen, or the learnedest physicians, that are unacquainted with the nature and properties of the centre of gravity, and the several kinds of levers, the wedge, etc.

III. Nay, there are several doctrines about physical things, that cannot be well explicated, and some of them not perhaps so much as understood, without mechanicks.

That, which emboldens me to propose a thing, seems so paradoxical, is, that there are many phænomena of nature, whereof though the physical causes belong to the consideration of the naturalist, and may be rendered by him: yet he cannot rightly and skillfully give them without taking in the causes statical, hydrostatical, &c. (if I may so name them) of those phænomena, i.e. such instances as depend upon the knowledge of mechanical principles and disciplines.

Of this we have an obvious example in that familiar observation, that we partly touched upon just now about the swimming and sinking of wood in water. For if it be demanded, why wood does rather swim upon water than sink to the bottom of it, a school-philosopher would answer, that wood abounds with air, which being an element very much lighter than water, keeps it aloft upon the surface of that liquor. But this answer will scarce satisfy a naturalist versed in hydrostaticks. For not now to question what is taken for granted, that there is a positive levity, and that the air is endowed with that quality, experience shews us, that though when wood is not heavier than so much water, as is equal to it in bulk, it will swim; yet in case it be heavier than so much water, it will sink. As we see in divers woods, and particularly in guaicaum, which I therefore the rather name, because chemists observe, that if it be burnt, it leaves far less ashes (and such are supposed to contain the terrestrial and heavy parts) behind it, than many woods, that we know will float in water. And though stones and iron be, upon the score of their weight, believed to be bodies, that have little air in them, yet if the liquor, into which they are put, be heavier, bulk for bulk, than they, they will not sink but float, and if forcibly depressed, they will emerge; as you may try, when you please, by putting stones or iron, or the like ponderous body upon quick-silver, or melted lead; so that we need not here consider, whether air be, or be not predominant in a proposed body, when we would know, whether it will, or will not sink in an assigned liquor.

And though we should admit the air, whether included in the pores, or looked upon as an elementary principle, to be the cause of its being lighter than an equal bulk of liquor, yet the air would be but the remote cause of its swimming, its immediate cause being, that the floating body is lighter than an equal bulk of the liquor, and therefore the same body, without acquiring or loosing air, may swim in one kind of water, and sink in another. As in the case of heavy bodies, as loaden ships, that having prosperously sailed over the sea, are recorded to have sunk as soon as they come into harbour, i.e. into a more fresh water; and an egg, that will sink in common water, will swim in a strong brine. Nay a body may (as I, and others have tried) be so poised in water, that if the liquor be a little warmer, than when the body was poised in it, the body will sink; as it will emerge again upon the refrigeration of it.

And if this general answer of the lightness of the air will not give so good an account as hydrostatical principles, why a piece of wood will float or sink, it will much less give so satisfactory an account, why differing woods in the same water, or the same piece of wood in differing waters, will sink just so far, and no further, whereas, by hydrostatical principles, the phænomenon is easy to be accounted for, according to that theorem of *Archimedes*, ωεςὶ τῶν ὀχυμέναγ, that solids lighter than liquor they are put into, will sink in it so far, as that as much of the liquor as is equal in bulk to the demersed part, be equal in weight to the whole floating body: whence these corollaries are derived, that a floating body has the same proportion in weight to as much liquor as is equal to it in bulk, as the immersed part of the body has to the whole body. And likewise, that as much liquor, as is equal in bulk to the whole body, has the same proportion in weight to the said body, as the whole body has to that part of itself, which is beneath the surface of the liquor. And as these corollaries determine the proportion between the immersed and extant part of the floating body; so (to shew you, that these theories lead to practice) they suggest the way of making a small and light instrument, elsewhere described, to measure by a floating body the differing gravities of several liquors in reference to one another, as well as to the body itself . . .

. . . And it were easy to add a multitude of examples, whereof a good account will scarce be given by a naturalist, that is unacquainted with mechanicks, and may easily be assigned by one that is skilled in them. But referring the schoolmen to *Aristotle*'s mechanical questions, to shew them the necessity and usefulness of mechanical knowledge, to give the solution of sundry phænomena, that frequently occur, I will only add an example or two to make good the most paradoxical part of what I was saying; namely, that there are divers physico-mechanical phænomena, which are not to be I say not explicated, but so much as well understood, without the knowledge of mechnical disciplines.

There is a considerable theorem in hydrostaticks, which is thought to have been first taken notice of by *Mersennus*, and in a late writer, is thus expressed: *Velocitates motus aquæ descendentis & effluentis per tubos æqualium foraminum, sed inæqualium altitudinum, habent subduplicatam rationem altitudinum.* Of which the corollary is, that the tubes are in a duplicate ration to that of the velocities of the water, that subsides in, and runs out of them; so that to make one tube at a circular hole of the same diameter run out in the same time twice as much water as another, the greater ought to be not only twice, but four times as long as the shorter. And of the same proportion (my tryals about which I may elsewhere acquaint you with) divers other practical applications may be made, which must not be here insisted on.

IV. As I formerly said of the mathematicks, so I now say of the mechanicks, that they may assist the naturalist to multiply experiments by those enquiries, that they will suggest, and those inferences and applications, whereto they may lead us.

Of this we have a noble instance in the great variety of tryals, which enquiries, versed in hydrostaticks, and other mechanical disciplines, have upon the score of their being so qualified, been either prompted, or at least assisted to make, about the famous quicksilver-experiment devised by *Torricellius*; about which though so much has been done already, yet almost every year brings forth new phænomena.

Another example to our present purpose we may take from the great number of new propositions, that the diligent *Mersennus* has given us in his balisticks, about the force and effects of bows, and the like springy bodies. But a yet more noble instance is given us by the most ingenious *Galilæo*, who, as we may learn from the already mentioned French writer, that has given us an account of *Galilæo*'s new thoughts in that language, has published so many propositions (of which he sets down 19 or 20, with the demonstrations) about the resistance of bodies to be broken, and the weights requisite to break them, and the lengths, at which they may be broken by their own weight, that he has reduced them into the form, and given them the title of a new art.

To all which I shall need to add no more, than that he, who knows and considers, what a variety of useful propositions have been, or may be mechanically deduced from the observation of *Archimedes*, that a solid body weighs less in water than in the air, by the weight of water equal in bulk to that body, will easily dispense with me for not adding any farther instances on this occasion . . .

V. Besides the utilities, that may be ascribed to the mechanicks in common, with the more speculative mathematical disciplines, they have some, as I formerly intimated, that are more peculiarly their own, since they may be of great use to the naturalist in making of such instruments and tools, as for many of his observations, trials, and other purposes, he may either absolutely need, or advantageously imploy.

Of this we have an example in the mariner's compass, as it is called; which is so necessary to those remote navigations, whereto natural philosophy and mankind owes so much. For though *Baptista Porta* does, as well as other authors, ascribe the invention of the directive faculty of the magnetick needle to one of his country-men (*Amalphi*, in the kingdom of *Naples*,) yet he confesses, that for want of the knowledge of making such sea-compasses as we now use, this lucky inventor was fain to make use of a piece of wood or straw, to keep the needle a float, and then imbue it with a magnetick vertue; which was a shift subject to great and manifest inconveniences. And indeed, notwithstanding the knowledge of the verticity of magnetical needles, if by that of the properties of the center of gravity, or some practices derived thence, some men, versed in mechanicks, had not devised a way so to poise the needle, that notwithstanding the rolling and tossing of the ship, it will continue horizontal enough to direct the pilot; what would become of him in those storms, when he has most need of a faithful guide? . . .

By the help of small valves, and the knowledge of the spring of compressed air, have been made those wind-guns, which may be employed, not only to weigh the air, (whose weight we found them to evince, but not determine), but to kill deer, and other game, without making a great noise, that would fright away the rest . . .

The concluding intimation I mean to give you, is, that I have not hitherto mentioned a service, that mathematicks and mechanicks may often do the naturalist, which is not fit to be silently pretermitted; and it is, that by lineal schemes, pictures, and instruments, they may much assist the imagination to conceive many things, and thereby the understanding to judge of them, and deduce new contrivances from them.

SOURCE: T. Birch (ed). *Boyle's Works*, vol III (1744), 162–6, 'Of the Usefulness of Mechanical Disciplines to Natural Philosophy'

THOMAS SPRAT, 1635–1713

Thomas Sprat was one of several high-ranking clerics who was also a member of the Royal Society. Himself to become Bishop of Rochester, he took a lively interest in the proceedings of the early experimenters, and soon began the work of writing a history and defence of the Society's undertakings. The early 1660s were years when the Society attracted a fashionable following of courtiers and aristocrats as patrons and wits inspired in some measure by the King's own interest, but when it also attracted some pusillanimous critics and pamphleteers sure that it was perverting youth, learning, the universities, morals, and religion, Sprat set out to defend it from all these accusations.

The fashionable interest led by men like Brouncker had to compete with the antagonistic views of Henry Stubbe, whose disgruntlement probably stemmed from his own exclusion from the Society. Stubbe and his minions argued that the idea of learning represented by the Royal Society was one hostile to social order and to the established canons of learning embodied in the universities and the church. Stubbe felt that any attack upon these canons of accepted educational practice would lead to decay in morals and in the state: he could persuade many people that the corruption and licence of Charles's court had something to do with the new philosophy that they were patronising.

Sprat in the extract below (11) defends the Royal Society against the charge that it seeks to overthrow all classical learning, to deny the Ancients their proper place, and to introduce novelty for no sake but its own. Sprat sees the Moderns as continuing the Ancients' true spirit, the spirit of independent reasoned enquiry which had been lost in centuries of textual obscurities and blind reverence to authority. Sprat accepted that classical and authoritarian didacticism should hold sway in schools and colleges teaching young men; but for men of more mature understanding Sprat recommends the freedom of enquiry into natural phenomena which the Royal Society to him embodies. The Royal Society, he argues, is nearer to the spirit and therefore to the wishes of the Ancients than was the atmosphere of the deferential schoolroom passing comment on the works of Aristotle or Plato. Sprat tries to balance his acceptance of the need for social order and authority in education with his belief in the vitality and the importance of the Royal Society endeavour, in bringing learning to gentlemen, and in leading to the advancement of understanding.

11 Thomas Sprat
'DEFENCE OF THE ROYAL SOCIETY'

But now it being a fit time to stop, and breathe a while, and to take a review of the ground, that we have pass'd. It will be here needful for me, to make an Apology for myself, in a matter, which, if it be not before-hand remov'd, may chance to be very prejudicial to mens good opinion of the *Royal Society* it self, as well as of its *Historian*. I fear, that this *Assembly* will receive disadvantage enough, from my weak management of their cause, in many other particulars: so that I must not leave them, and myself unjustify'd, in this wherein we have so much right on our sides. I doubt not then, but it will come into the thoughts of many *Criticks*, (of whom the World is now full) to urge against us, that I have spoken a little too sparingly of the Merits of former Ages; and that this Design seems to be promoted, with a malicious intention of disgracing the Merits of the *Antients*.

But First, I shall beseech them, calmly to consider; whether they themselves do not more injure those great Men, whom they would make the Masters of our Judgments, by attributing all things to them so absolutely; then we, who do them all the Justice we can, without adoring them? It is always esteem'd the greatest mischief, a man can do those whom he loves, to raise mens expectations of them too high, by undue, and impertinent commendations. For thereby not only their enemies, but indifferent men, will be secretly inclin'd to be more watchful over their failings, and to conspire in beating down their Fame. What then can be more dangerous to the honor of Antiquity; then to set its value at such a rate, and to extol it so extravagantly, that it can never be able to bear the tryal, not onely of envious, but even of impartial Judges? It is natural to Mens minds, when they perceive others to arrogate more to themselves, then is their share; to deny them even that, which else they would confess to be their Right. And of the Truth of this, we have an instance of far greater concernment, then that which is before us. And that is, in *Religion* it self. For while the *Bishops* of *Rome* did assume an infallibility, and a sovereign Dominion over our Faith: the *reformed Churches* did not onely justly refuse to grant them that, but some of them thought themselves oblig'd to forbear all communion with them, and would not give them that respect, which possibly might belong to so antient, and so famous a *Church*; and which might still have been allowed it, without any danger of Superstition.

But to carry this Dispute a little farther: What is this, of which they accuse us? They charge us with immodesty in neglecting the guidance of wiser, and more discerning Men, then our selves. But is not this rather the greatest sign of Modesty, to confess, that we our selves may err, and all mankind besides? To acknowledge the difficulties of Science? and to submit our minds, to all the least Works of Nature? What kind of behavior do they exact from us in this case? That we should reverence the Footsteps of *Antiquity*? We do it most unanimously. That we should subscribe to their sense, before our own? We are willing, in probabilities; but we cannot, in matters of Fact: for in them we follow the most antient Author of all others, even *Nature* it self. Would they have us make our eies behold things, at no farther distance, than they saw? That is impossible; seeing we have the advantage of standing upon their shoulders. They say, it is insolence, to prefer our own inventions before those of our *Ancestors*. But do not even they the very same things themselves, in all the petty matters of life? In the Arts of War, and Government; In the making, and abolishing of Laws; nay even in the fashion of their Cloaths, they differ from them, as their humour, or Fancy leads them. We approach the Antients, as we behold their Tombs, with veneration: but we would not therefore be confin'd to live in them altogether: nor would (I believe) any of those, who profess to be moste addicted to their Memories. They tell us, that in this corruption of Manners, and sloth of Mens Minds, we cannot go beyond those, who search'd so diligently, and concluded so warily before us. But in this they are confuted by every days experience. They object to us *Tradition*, and the consent of all Ages. But do we not yet know the deceitfulness of such Words? Is any man, that is acquainted with the craft of founding *Sects*, or of managing Votes in *popular Assemblies*, ignorant, how easie it is to carry things in a violent stream? And when an opinion has once master'd its first opposers, and setled it self in Mens Passions, or Interests: how few there be, that coldly consider, what they admit for a long time after? So that when they say, that *all Antiquity* is against us; 'tis true, in shew, they object to us, the Wisdom of many Ages; but in reality, they onely confront us, with the Authority of a few leading Men. Nay, what if I should say, that this honor for the dead, which such men pretend to, is rather a worshiping of themselves, than of the *Antients*? It may be well prov'd, that they are more in love with their own *commentaries*, then with the *Texts* of those, whom they seem to make their Oracles: and that they chiefly doat on those Theories, which they themselves have drawn from them: which, it is likely, are almost as far distant from the Original meaning of their Authors, as the Positions of the *New Philosophers* themselves.

But to conclude this Argument (for I am weary of walking in a rode so trodden) I think I am able to confute such men by the practice of those very Antients, to whom they stoop so low. Did not they trust themselves, and their own Reasons? Did not they busie themselves in inquiry, make new Arts, establish new

Tenents, overthrow the old, and order all things as they pleas'd, without any servile Regard to their Predecessors? The *Grecians* all, or the greatest part of them, fetch'd their Learning from *Egypt*. And did they blindly assent to all, that was taught them by the *Priests* of *Isis*, and *Osiris*? If so; then why did they not, together with their Arts, receive all the infinit Idolatries, which their Masters embrac'd? seeing it is not to be question'd, but the *Egyptians* deliver'd the rites of their Religion to strangers, with as much Solemnity at least, as they did the Mysteries of their *Hieroglyphicks* or *Philosophy*. Now then, let *Pythagoras*, *Plato*, and *Aristotle*, and the rest of their wise Men, be our examples, and we are safe. When they travell'd into the *East*, they collected what was fit for their purpose, and suitable to the Genius of their Country; and left the superfluities behind them: They brought home some of their useful Secrets: but still counted their worshiping a Dog, or an Onion, a Cat, or a Crocodile, ridiculous. And why shall not we be allow'd the same liberty, to distinguish, and choose, what we will follow? Especially, feeling in this, they had a more certain way of being instructed by their Teachers, then we have by them: They were present on the place: They learn'd from the Men themselves, by word of mouth; and so were in a likely course to apprehend all their Precepts aright: whereas we are to take their Doctrines, so many hundred years after their death, from their Books only, where they are for the most part so obscurely express'd, that they are scarce sufficiently understood by the *Grammarians*; and *Linguists* themselves, much less by the *Philosophers*.

In few words therefore, let such men believe, that we have no thought of detracting from what was good in former times: But, on the contrary, we have a mind to bestow on them, a solid praise, instead of a great, and an empty. While we are raising new Observations upon Nature, we mean not to abolish the Old, which were well, and judiciously establish'd by them: No more, then a *King*, when he makes a new Coyn of his own, does presently call in that, which bears the Image of his Father: he onely intends thereby to increase the current Money of his

Kingdom, and still permits the one to pass, as well as the other. It is probable enough, that upon a fresh survey, we may find many things true, which they have before asserted: and then will not they receive a greater confirmation, from this our new and severe approbation, then from those men, who resign up their opinions to their Words only? It is the best way of honoring them, to separate the certain things in them, from the doubtful: For that shews, we are not so much carri'd towards them, by rash affection, as by an unbyass'd Judgement. If we would do them the most right; it is not necessary we should be perfectly like them in all things. There are two principal Ways of preserving the Names of those, that are pass'd: the one, by *Pictures*; the other, by *Children*: The *Pictures* may be so made, that they may far neerer resemble the Original, then Children do their Parents: and yet all Mankind choose rather to keep themselves alive by Children, then by the other. It is best for the *Philosophers* of this Age to imitate the *Antients* as their *Children*: to have their blood deriv'd down to them; but to add a new Complexion, and Life of their own: While those, that indeavour to come neer them in every Line, and Feature, may rather be call'd their dead *Pictures*, or *Statues*, then their *Genuine Off-Spring*.

SOURCE: Thomas Sprat. *The History of the Royal Society* (1677), 46–51

SIR ISAAC NEWTON, 1642–1727

Sir Isaac Newton, the Cambridge mathematician, thought his purpose in life was to bring enlightenment to his generation. Most of his generation came to agree, for in his sound and imposing Principia Mathematica *Newton laid down the necessary proofs for the long discovered propositions concerning planetary movements, and demonstrated finally the primacy of the mathematical method in treating natural philosophy in all its mechanical branches. Newton was also long remembered for the provoking and impressionistic questions set out in the* Optics, *which were to influence the direction of enquiry for eighteenth-century science.*

Newton's view in the Preface to the first edition (12) is a mechanistic one. He argues that the natural philosopher should perceive the motions which instigate the forces of nature, and then from these forces the philosopher should demonstrate the phenomena of the universe. Newton acknowledged the debt he owed to Edmund Halley, the Royal Society astronomer, who had provoked and persuaded Newton into writing down and publishing his demonstrations, proofs, and lemmae. Without Halley's persuasion Newton's Principia *might never have been written or printed. In the Introduction to Book III (13) Newton explains why his work is so complicated. Fearful of disputes, afraid lest people should misconstrue his vision, or afraid that a little learning would make everyone conversant with the rudimentary elements of his world system so that they could criticise it, Newton told his readers that he had therefore couched it all in mathematical language. The system of the world explained in Book III depended upon the mathematical propositions concerning force and motion in the preceding two books. But Newton relented and advised his readers that it would be sufficient to read the definitions and Laws of Motion early on in the first book and then pass on to understand the system of the world explained in Book III.*

Newton's rules of reasoning in philosophy (14) were much discussed and debated. Few believed that he ever himself strictly kept to them, although he always attempted to maintain that he avoided hypothesising. He argued that a work like the Optics *was merely a set of queries, his own opinions culled from insufficient mathematics or evidence, which were to be kept as questions to avoid masquerading as serious proven scientific work. The importance of Newton's rules of reasoning lies in the assertion that like causes produce like effects wheresoever they occur in the universe. It was this premise which permitted the Newtonian world picture to take shape in the form it did.*

12 Sir Isaac Newton
'PRINCIPIA MATHEMATICA': PREFACE TO FIRST EDITION

SINCE the ancients (as we are told by Pappus) made great account of the science of Mechanics in the investigation of natural things; and the moderns, laying aside substantial forms and occult qualities, have endeavoured to subject the phænomena of nature to the laws of mathematics; I have in this treatise cultivated Mathematics, so far as it regards Philosophy. The ancients considered Mechanics in a twofold respect; as rational, which proceeds accurately by demonstration, and practical. To practical Mechanics all the manual arts belong, from which Mechanics took its name. But as artificers do not work with perfect accuracy, it comes to pass that Mechanics is so distinguished from Geometry, that what is perfectly accurate is called Geometrical, what is less so is called Mechanical. But the errors are not in the art, but in the artificers. He that works with less accuracy, is an imperfect Mechanic, and if any could work with perfect accuracy, he would be the most perfect Mechanic of all. For the description of right lines and circles, upon which Geometry is founded, belongs to Mechanics. Geometry does not teach us to draw these lines, but requires them to be drawn. For it requires that the learner should first be taught to describe these accurately, before he enters upon Geometry; then it shews how by these operations problems may be solved. To describe right lines and circles are problems, but not geometrical problems. The solution of these problems is required from Mechanics; and by Geometry the use of them, when so solved, is shewn. And it is the glory of Geometry that from those few principles, fetched from without, it is able to produce so many things. Therefore Geometry is founded in mechanical practice, and is nothing but that part of universal Mechanics which accurately proposes and demonstrates the art of measuring. But since the manual arts are chiefly conversant in the moving of bodies, it comes to pass that Geometry is commonly referred to their magnitudes, and Mechanics to their motion. In this sense Rational Mechanics will be the science of motions resulting from any forces whatsoever and of the forces required to produce any motions, accurately proposed and demonstrated. This part of Mechanics was cultivated by the ancients in the Five Powers which relate to manual arts, who considered gravity (it not being a manual power) no otherwise than as it moved weights by those powers. Our design not respecting arts but philosophy, and our subject, not manual but natural powers, we consider chiefly those things which relate to gravity, levity, elastic force, the resistance of fluids, and the like forces whether attractive or impulsive. And therefore we offer this work as mathematical principles of philosophy. For all the difficulty of philosophy seems to consist in this from the phænomena of motions to investigate the forces of Nature, and then from these forces to demonstrate the other phænomena. And to this end, the general propositions in the first and second book are directed. In the third book we give an example of this in the explication of the System of the World. For by the propositions mathematically demonstrated in the first books, we there derive from the celestial phænomena, the forces of Gravity with which bodies tend to the Sun and the several Planets. Then from these forces by other propositions, which are also mathematical, we deduce the motions of the Planets, the Comets, the Moon, and the Sea. I wish we could derive the rest of the phænomena of Nature by the same kind of reasoning from mechanical principles. For I am induced by many reasons to suspect that they may all depend upon certain forces by which the particles of bodies, by some causes hitherto unknown, are either mutually impelled towards each other and cohere in regular figures, or are repelled and recede from each other; which forces being unknown, Philosophers have hitherto attempted the search of Nature in vain. But I hope the principles here laid down will afford some light either to that, or some truer, method of Philosophy.

In the publication of this Work, the most acute and universally learned Mr. Edmund Halley not only assisted me with his pains in correcting the press and taking care of the Schemes, but it was to his solicitations that its becoming publick is owing. For when he had obtained of me my demonstrations of the figure of the celestial orbits, he continually pressed me to communicate the same to the Royal Society; who afterwards by their kind encouragement and entreaties, engaged me to think of publishing them. But after I had begun to consider the inequalities of the lunar motions, and had entered upon some other things relating to the laws and measures of gravity, and

E   65

other forces; and the figures that would be described by bodies attracted according to given laws; and the motion of several bodies moving among themselves; the motion of bodies in resisting mediums; the forces, densities, and motions of mediums; the orbits of the Comets, and such like; I put off that publication till I had made a search into those matters, and could put out the whole together. What relates to the Lunar motions (being imperfect) I have put all together in the corollaries of prop. 66. to avoid being obliged to propose and distinctly demonstrate the several things there contained in a method more prolix than the subject deserved, and interrupt the series of the several propositions. Some things found out after the rest, I chose to insert in places less suitable, rather than change the number of the propositions and the citations. I heartily beg that what I have here done may be read with candour, and that the defects I have been guilty of upon this difficult subject may be, not so much reprehended, as kindly supplied, and investigated by new endeavours of my readers.

SOURCE: Sir Isaac Newton. *The Mathematical Principles of Natural Philosophy*, translated by A. Motte, vol I (1729)

13 Sir Isaac Newton
'PRINCIPIA MATHEMATICA': INTRODUCTION
TO BOOK III

IN the preceding books I have laid down the principles of philosophy; principles not philosophical, but mathematical; such, to wit, as we may build our reasonings upon in philosophical enquiries. These principles are, the laws and conditions of certain motions, and powers or forces, which chiefly have respect to philosophy. But lest they should have appeared of themselves dry and barren, I have illustrated them here and there, with some philosophical scholiums, giving an account of such things, as are of more general nature, and which philosophy seems chiefly to be founded on; such as the density and the resistance of bodies, spaces void of all bodies, and the motion of light and sounds. It remains, that from the same principles, I now demonstrate the frame of the System of the World. Upon this subject, I had indeed compos'd the third book in a popular method, that it might be read by many. But afterwards considering that such as had not sufficiently enter'd into the principles, could not easily discern the strength of the consequences, nor lay aside the prejudices to which they had been many years accustomed; therefore to prevent the disputes which might be rais'd upon such accounts, I chose to reduce the substance of that book into the form of propositions (in the mathematical way) which should be read by those only, who had first made themselves matters of the principles establish'd in the preceding books. Not that I would advise any one to the previous study of every proposition of those books. For they abound with such as might cost too much time, even to readers of good mathematical learning. It is enough if one carefully reads the definitions, the laws of motion, and the first three sections of the first book. He may then pass on to this book of the System of the World, and consult such of the remaining propositions of the first two books, as the references in this, and his occasions, shall require.

SOURCE: as Document 12, vol II, 200–1

14 Sir Isaac Newton
'PRINCIPIA MATHEMATICA':
THE RULES OF REASONING IN PHILOSOPHY

Rule I

We are to admit no more causes of natural things, than such as are both true and sufficient to explain their appearances.

To this purpose the philosophers say, that Nature do's nothing in vain, and more is in vain, when less will serve; For Nature is pleas'd with simplicity, and affects not the pomp of superfluous causes.

Rule II

Therefore to the same natural effects we must, as far as possible, assign the same causes.

As to respiration in a man, and in a beast; the descent of stones in Europe and in America; the light of our culinary fire and of the Sun; the reflection of light in the Earth, and in the Planets.

Rule III

The qualities of bodies, which admit neither intrusion nor remission of degrees, and which are found to belong to all bodies within the reach of our experiments, are to be esteemed the universal qualities of all bodies whatsoever.

For since the qualities of bodies are only known to us by experiments, we are to hold for universal, all such as universally agree with experiments; and such as are not liable to diminution, can never be quite taken away. We are certainly not to relinquish the evidence of experiments for the sake of dreams and vain fictions of our own devising; nor are we to recede from the analogy of Nature, which uses to be simple, and always consonant to it self. We no other ways know the extension of bodies, than by our senses, nor do these reach it in all bodies; but because we perceive extension in all that are sensible, therefore we ascribe it universally to all others also. That abundance of bodies are hard we learn by experience. And because the hardness of the whole arises from the hardness of the parts, we therefore justly infer the hardness of the undivided particles not only of the bodies we feel but of all others. That all bodies are impenetrable, we gather not from reason, but from sensation. The bodies which we handle we find impenetrable, and thence conclude impenetrability to be an universal property of all bodies whatsoever. That all bodies are moveable, and endow'd with certain powers (which we call the *vires inertiæ*) of persevering in their motion or in their rest, we only infer from the like properties observ'd in the bodies which we have seen. The extension, hardness, impenetrability, mobility, and *vis inertiæ* of the whole, result from the extension hardness, impenetrability, mobility and *vires inertiæ* of the parts: and thence we conclude the least particles of all bodies to be also all extended, and hard and impenetrable, and moveable, and endow'd with their proper *vires inertiæ*. And this is the foundation of all philosophy. Moreover, that the divided but contiguous particles of bodies may be separated from one another, is a matter of observation; and, in the particles that remain undivided, our minds are able to distinguish yet lesser parts, as is mathematically demonstrated. But whether the parts so distinguish'd, and not yet divided, may, by the powers of nature, be actually divided and separated from one another, we cannot certainly determine. Yet had we the proof of but one experiment, that any undivided particle, in breaking a hard and solid body, suffer'd a division, we might by virtue of this rule, conclude, that the undivided as well as the divided particles, may be divided and actually separated to infinity.

Lastly, If it universally appears, by experiments and astronomical observations, that all bodies about the Earth gravitate towards the Earth; and that in proportion to the quantity of matter which they severally contain; that the Moon likewise, according to the quantity of its matter, gravitates towards the Earth, that on the other hand our Sea gravitates towards the Moon; and all the Planets mutually one towards another; and the Comets in like manner towards the Sun; we must in consequence of this rule, universally allow, that all bodies whatsoever are endow'd with a principle of mutual gravitation. For the argument from the appearances concludes with more force for the universal gravitation of all bodies, than for their impenetrability; of which among those in the celestial regions, we have no experiments, nor any manner of observation. Not that I affirm gravity to be essential to bodies. By their *vis insita* I mean nothing but their *vis inertia*, This is immutable. Their gravity is diminished as they recede from the Earth.

Rule IV

In experimental philosophy we are to look upon propositions collected by general induction from phænomena as accurately or very nearly true, notwithstanding any contrary hypotheses that may be imagined, till such time as other phænomena occur, by which they may either be made more accurate, or liable to exceptions.

This rule we must follow that the argument of induction may not be evaded by hypotheses.

SOURCE: as Document 12, vol II, 202–5

Part Two

ENTERPRISE AND ACHIEVEMENT

The seventeenth century was punctuated not only by endless and some-times futile debates about method and the philosophy of science but also by many startling discoveries and theories. The century was one of great endeavour in the natural sciences: a period of intense and intensifying activity, a period of great intellectual passions. There were those who saw in the discoveries of Galileo, Harvey, Boyle and Newton proof that scientific advance took place through the advances made by great men: proof that crucial experiments like those alleged to have taken place from the Leaning Tower of Pisa, where Galileo was reputed to have dropped two objects to verify statements about free fall, or Pascal's demonstration of the Torricellian vacuum by taking a column of mercury up the Puy de Dome, could enlighten their generation. These men felt that after Pascal's ingenious journey, demonstrating that the column of mercury contained in a glass tube did rise higher up the tube the higher up the mountain you took it, it had been proved conclusively that air pressure exerted a force upon the column of mercury and proved that there was a vacuum above the top of the mercury level. Others thought of achievement as being the collective and collaborative endeavours of hosts of lesser and greater men together, assimilating information, trying and failing, and then trying again. But most seventeenth-century men thought in terms of great projects and great advances; their successes and failures are eloquent testimony to their essays in scientific advance.

Wherever you looked in Western Europe in the century, with the ex-ception of closed and backward Spain, there were men devoted to the cause of natural philosophy, pressing ahead with some programme of enquiry. There was the Northumberland circle in England at the beginning of the century, interested in alchemy and the atomic philosophy. In London the

71

Professors of Gresham College were studying the arts of navigation, surveying and mathematics. In nearby France men like Descartes and Mersenne were beginning their education, while in Italy Galileo was soon to take up the neo-platonic enquiries of his father, moving from music to astronomy as his principal study. In Florence and Rome groups of men were already meeting to discuss scientific and literary matters, and at Padua the great School of Anatomy was pushing forward with its studies.

Some 30 years later from Great Tew, where an important circle of English intellectuals were meeting, to Paris, and from Florence to Padua, the pattern was similar. At Gresham College, and Leyden and other Dutch centres, artisans and engineers were applying their talents and the knowledge culled from empirical progress to aid the cause of scientific advance.

By the end of the century these beginnings had produced a scientific mania; it had spread into Germany and the Northern Baltic, it had flourished in Leipzig, and in the Acta Eruditorum it had penetrated the country parishes of rural England and the seminaries of Catholic France. Academies and journals had been founded, the whole air of Europe was affected by the vogue for natural philosophy. The movement had discovered the truth of the old adage that nothing succeeds like success.

What had the successes been? They had been so many that they had influenced the way every intelligent person looked at the world of the microcosm and the macrocosm. The astronomers had firmly placed the earth in revolution around the sun, and saw the sun as one among many in the world system. The earth was displaced from the centre, and the new mathematical logic provided formulae concerning planetary orbits that satisfactorily explained the motions of the near planets. The physicists had made many discoveries about matter and motion and had seen their work finally integrated with astronomy by Newton. Their understanding of force, movement, free fall and dynamics aided the progress of mechanistic science. The philosophers saw the work of Descartes, Locke and Leibniz, among others, lead to a fundamental revision of the intellectual pillars of the universe. The Cartesian dualism and the Leibnizian Monadology substantially altered the way in which men conceived of their worlds.

The chemists had strenuously tried to transmute metals into gold and to find the elixir of eternal life; but they had also produced Boyle's work on chemical indicators and had clarified in part the nature of elements and compounds; while there was a healthy scepticism about the old four elements theory in common currency. The medical men had worked hard and assiduously during the century, like everyone else. By its end there was a far more detailed anatomical science, and the range of medicine had been much extended by the iatro-chemical school. Above all, doctors now knew of the circulation of the blood, and had by the end of the century wisely abandoned attempts at primitive blood transfusions, which had been brought into disrepute by a death in Paris. They had been based on an ignorance of blood groupings.

The mathematicians during the century had seen the range of their discipline widely extended, with the addition of logarithms and other forms of computational aid, and with the successful integration of mathematics into the whole realm of natural philosophical enquiry. Newton and Leibniz between them had pioneered calculus, some progress had been made with developing Euclidean geometry, and the Cartesian coordinate geometry represented a great advance when graphs could be transmuted into equations and vice versa with comparative ease.

Botanists had produced a more detailed catalogue of the world's wonders than the Ancients and their sixteenth-century forbears had bequeathed them. It was every gentleman's task to travel round the countryside cataloguing plants and publishing details of the rarities of his own locality. Beneath the presiding genius of John Ray and others with a taxonomical bent of mind, large histories of natural phenomena were constructed. Microscopists had similarly penetrated the inner recesses of small animals and plants and discovered the secrets of a world which was closed to anyone before Hooke, Power and Leeuwenhoek had perfected their various optical instruments.

It would be false, however, to give a picture of unrelieved progress towards enlightenment. Many of these well known achievements were less significant than contemporaries or some later historians have imagined. Harvey may well have discovered the circulation of the blood, but it had no important practical effect on medicine; chemists had no alternative theoretical structure to turn to if they agreed with Boyle that the old four elements and humours theory was inadequate. Not everyone immediately accepted the Newtonian picture of the universe, though many of the endless disputes were not created by the ideas and perseverance of

*misguided conservatives standing out against brilliant scientific geniuses.
For example, those who denied the existence of the vacuum were by no
means stupid: they were right to suggest that vapour filled the Toricellian
vacuum at the head of the mercury column in an inverted glass tube,
correct to suggest that leakage occurred, and not stupid to assert that perhaps nature abhors a vacuum. Similarly, those who protested against
Galileo's findings did have serious arguments to put forward, given the
limitations of glass grinding in the seventeenth century, and the problems
of producing a lens capable of seeing the heavens clearly.*

*Any study of the scientific movement of the seventeenth century should
also remember that the century which produced Newton's work on gravity
also produced his work on alchemy, prophecy and chemistry. The century
that produced Galileo also produced Fludd, a strange mystic iatrochemist, and Paracelsus. The century that saw the discovery of calculus
also spent far more time and trouble discussing the mystical significances
of numbers, the nature of predictions and horoscopes, and the magical
powers of matter. It is this fascinating diversity of experience which
animates the intellectual life of the seventeenth century. We should resist
the prefaces of men who have come to be highly regarded, and penetrate
their propaganda, which strains to see two parties, one of rational
mathematical progressives and the other of clerical obstructionists. We
should instead see many men all over Western Europe struggling for
achievement, borrowing here, there and everywhere, some of whom
managed to exert a greater and more long-lived influence than others.*

PHYSICS AND MATHEMATICS

*From the logarithms of Napier to the calculus of Newton and Leibniz,
from the early days of Gresham College in London to the sophisticated
mathematics of the later Académie des Sciences, the seventeenth century
was above all a century of mathematical advance. Mathematics was the
new language, the secret of nature, the root of physics, and physics the
basis of the universe.*

*It was not only the age of Descartes and Roberval, the great French
mathematicians, nor of Napier and Leibniz; while the significance of
their work was great, advancing coordinate geometry, logarithmic calculation, and producing calculus itself, its sophistication meant that it only
penetrated to a small circle of illuminati. There was an attempt by some*

men in the seventeenth century to make mathematics more readily available and we can see in the extract below (*15*) a translation by Edward and Samuel Wright that made Napier's work more easily accessible. There were several editions of Euclid published during the century to make Euclid both more easily and readily obtainable for the average reader, of which the De Chales volume is but one example (*16*).

Mathematics also became a form of popular diversion. Wilkins's Mathematical Magick was perhaps the most famous of a genre which showed the practical applications and diversions that could come from mechanics and mathematics (*17*). An understanding of mechanics led men to produce all manner of ingenious machines and devices that became the toys of the late seventeenth- and early eighteenth-century drawing rooms.

The contribution mathematics made to the serious physics of the century is everywhere manifest. It was particularly true in the field of motion and kinematics, pioneered and developed by Galileo and his followers; it was also true of dioptrics and optics, seen at its most obvious in Newton but also apparent in Molyneux's work (*18*).

As mathematics developed, there seemed nothing it could not solve. The seventeenth-century enquirers discussed the great problems of infinites and infinitesimals, of squaring the circle and analysing graphs, and saw mathematics constantly within the larger perspective of philosophy and natural science. Mathematicians like Ward, an Englishman and Oxford professor, may well have specialised, but they were also thoroughly aware of the wider ramifications of their discipline in fostering the whole scientific movement of their epoch.

15 John Napier
'LOGARITHMS': DEDICATION AND AUTHOR'S PREFACE TO ENGLISH TRANSLATION

SAMUEL WRIGHT'S DEDICATION TO THE EAST INDIA COMPANY

Your favours towards my deceased Father, and your imployment of him in businesse of this nature, but chiefly your continuall imployment of so many Mariners in so many goodly and costly ships, in long and dangerous voyages, for whose use (though many other wayes profitable) this little booke is chiefly behoovefull: may chalenge an interest in these his labours. This *Book* is noble by birth, as being descended from a Noble Parent, & not ignoble by educatiõ, having learned to speake English of my late Father, a man in the judgment of the learned, and experience of the common sort, famous for knowledge and practise in the Mathematickes: whose care thereof was so great, to send it abroad with the true resemblance of his worthy father, and sufficient knowledge of the English tongue to instruct our Countrey-men, that hee procured the Authors perusall of it: who after great paines taken therein, gave approbation to it, both in substance and forme, as now I present it unto you. I am the bolder thus to do, in regard it is not unknowne to many men, that my said father spent a great part of his time in study of the Art of Navigation, and had gathered much understanding by his owne practise in some voyages to sea with the right Honourable the Earle of *Cumberland* deceased: whereupon he published a painful worke, discovering errours committed by Mariners in that Art, with corrections and ready wayes for reformation thereof. So that I thinke it is out of doubt, that his judgement therein was great. And seeing hee not onely gave much commendation of this worke (and often in my hearing) as of very great use for Mariners: but also to help the want of those that could not understand it in Latine, translated the same into English, and added thereto an instrumentall Table to finde the part proportional, whereof also the noble Authors approved well. I doubt not but it is apparent enough that he esteemed of it, and intended to have recommended it as a booke of more then ordinary worth, especially to Sea-men. But shortly after he had it returned out of *Scotland*, it pleased God to call him away afore he could publish it, or but write a description of the said instrumentall Table which he had devised, therefore hee left the publishing of it to me, as an inheritance, and the said description to his learned and kind friend Mr. *Henry Brigges*, who hath performed it accordingly. All which I humbly present unto you, hoping you shall receave as much profite by the use of it, as there hath been learning, care, and paines bestowed in the penning and fitting it thus to your hands.

Seeing there is nothing (right well beloved Students in the Mathematickes) that is so troublesome to Mathematicall practise, nor that doth more molest and hinder Calculators, then the Multiplications, Divisions, Square and cubical Extractions of great numbers, which besides the tedious expence of time, are for the most part subject to many slippery errors. I began therefore to consider in my minde, by what certaine and ready Art I might remove those hindrances. And having thought upon many things to this purpose, I found at length some excellent briefe rules to be treated of (perhaps) hereafter. But amongst all, none more profitable then this, which together with the hard and tedious Multiplications, Divisions, and Extractions of rootes, doth also cast away from the worke it selfe, even the very numbers themselves that are to be multiplied, divided and resolved into rootes, and putteth other numbers in their place, which performe as much as they can do, onely by Addition and Subtraction, Division by two, or Division by three: which secret invention, being (as all other good things are) so much the better as it shall be the more common, I thought good heretofore to set forth in Latine for the publique use of Mathematicians. But now some of our Countrey-men in this Island well affected to these studies, and the more publique good, procured a most learned Mathematician to translate the same into our vulgar English tongue, who after he had finished it, sent the Coppy of it to me, to bee seene and considered on by my selfe. I having most willingly and gladly done the same, finde it to bee most exact and precisely conformable to my minde and the originall. Therefore it may please you who are inclined to these studies, to receive it from me and the Translator, with as much good will as we recommend it unto you. Fare yee well.

SOURCE: E. Wright. *A Description of the Admirable Table of Logarithms* (London, 1616)

PREFACE

Having for a long time observ'd, that most of those, that take in hand the *Elements of Euclid*, are apt to dislike them, because they cannot presently discern, to what end those seemingly inconsiderable, and yet difficult Propositions, can conduce: I thought I should do an acceptable piece of service, in not only rendring them as easie as possible, but also adding to each proposition a brief account of some Use, that is made of them in the other parts of the Mathematicks. In prosecuting which design, I have been oblig'd to change some Demonstrations, that seem'd too intricate and perplex'd, and above the ordinary capacity of Beginners, and so substitute others more intelligible in their stead. For the same reason, I have demonstrated the fifth Book after a method much more clear, than that by *Equimultiples*, formerly used. I would not be thought to have set down all the Uses, that may be made of these Propositions: to have done that, would have oblig'd me to have compris'd the whole *Mathematicks* in this one Book; which would have render'd it both too large and too difficult. But I have contented my self with the choice of such, as may serve to point out some of the Advantages they afford us, and are also in themselves most clear, and most easie to be apprehended. I have distinguish'd them by *Inverted Commas*, that the Reader may know them; not desiring he should dwell too long upon them, or labour to understand them perfectly at first, since they depend on the Principles of the other Parts.

This therefore being the design of this small Treatise, I voluntary offer it to the publick, in an Age, whose *Genius* seems more addicted to the *Mathematicks*, than any that has preceded it.

Eight Books of the Elements of Euclid, *together with the Use of the* Propositions

'The design of Euclid in this Book is to lay down the First
Principles of Geometry; and to do it methodically, he begins
with Definitions, and the explication of the most ordinary
Terms. To these he adds some Postulata; and then, proposing
those known Maxims, in which natural reason does instruct us,
he pretends, not to advance a step farther, without a Demonstration, but to convince every man, even the most obstinate,
that will grant nothing, but what is extorted from him. In the
first Proposition, he treats of Lines, and the different Angles,
which are form'd by their concourse; and having occasion to
compare divers Triangles together, in order to demonstrate the
Properties of Angles, he makes that the business of the Eight
first Propositions. Then follow some Practical Instructions, how
to divide an Angle and a Line into two equal parts, and to
draw a Perpendicular. Next he shows the properties of a
Triangle, together with those of Parallel Lines; and having
thus finish'd the Explication of this first figure, he passes on to
Parallelograms, teaching the manner of reducing any Polygone,
or multangular figure into one more regular. Lastly, he finishes
the first Book with that famous Proposition of Pythagoras, *That
in every rectangular Triangle the Square of the * Base is equal to the
Squares of both the other Sides*.'

DEFINITIONS

1. *A Point is that which hath no parts.*

'This Definition must be understood in this sense: That
quantity, which we conceive without distinguishing its parts, or
so much as considering whether or no it has any, is a Mathematical point; which is therefore very different from those of
Zeno, which were suppos'd to be absolutely indivisible, and
therefore such, that we may reasonably doubt whether they are

* He calls that the Base, which is commonly call'd the Hypotenuse, i.e. the Line
that is opposite to the Right Angle.

possible; but the former we cannot doubt of, if we conceive them aright.'

2. *A Line is length without breadth*

'The sense of this definition is the same with the former, That quantity, which we conceive as length, without reflecting on its breadth or thickness, is that, which we understand by a Line; though it be impossible to draw a real Line, which will not be of a certain breadth. 'Tis commonly said, that a Line is produc'd by the motion of a Point; which ought to be carefully observ'd; for motion may on that manner produce any quantity whatsoever: but here, we must imagine a Point to be only so mov'd, as to leave one trace in the space, through which it passes, and then, that trace will be a line.'

3. *The two Extreams of a Line are Points*

4. *A right Line is that, whose points are equally plac'd between the two Extreams.*

'Or thus. A right Line is the shortest that can be drawn from one point to another. Or yet. The Extreams of a right Line may cast a shadow upon the whole Line.'

Source: C. F. Milliet De Chales. *The Elements of Euclid Explained, in a New but Most Easie Method* (English ed, Oxford, 1685), Preface and 1–3

17 John Wilkins
'THE BALANCE'

The first invention of the Ballance is commonly attributed to Astrea, who is therefore deified for the goddess of Justice; and that Instrument it self advanced amongst the Cœlestial signs.

The particulars concerning it, are so commonly known, and of such easie experiment, that they will not need any large explication. The chief end and purpose of it, is for the distinction of several ponderosities; For the understanding of which, we must note, that if the length of the sides in the Ballance, and the weights at the ends of them, be both mutually equal, then the Beam will be in a horizontal scituation. But on the contrary, if either the weights alone be equal, and not their distances, or the distances alone, and not the weights, then the Beam will accordingly decline.

As in this following diagram.

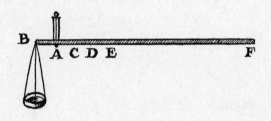

Suppose an equal weight at C, unto that at B, (which points are both equally distant from the center A,) it is evident that then the beam B F, will hang horizontally. But if the weight supposed at C, be unequal to that at B, or if there be an equal weight at D E, or any of the other unequal distances; the Beam must then necessarily decline.

With this kind of Ballance, it is usual by the help only of one weight, to measure sundry different gravities, whether more or less than that by which they are measured. As by the example here described, a man may with one pound alone, weigh any other body within ten pounds, because the heaviness of any weight doth increase proportionably to its distance from the Center. Thus one pound at D, will equiponderate unto two pounds at B, because the distance AD, is double unto AB. And for the same reason, one pound at E, will equiponderate to three pounds at B; and one pound at F, unto ten at B, because there is still the same disproportion betwixt their several distances.

This kind of Ballance is usually styled *Romana, statere*. It seems to be of ancient use, and is mentioned by Aristotle under the name of φάλαγξ.

Hence it is easie to apprehend how that false Ballance may be composed, so often condemned by the Wiseman, as being an abomination to the Lord. If the sides of the Beam be not eiqually divided, as suppose one have 10 parts, and the other 11, then any two weights that differ according to this proportion (the heavier being placed on the shorter side, and the lighter on the longer) will equiponderate. And yet both the scales being empty, shall hang in *equilibrio*, as if they were exactly just and true, as in this description.

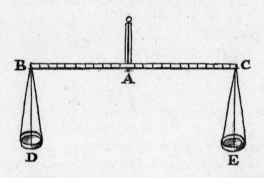

Suppose AC, to have 11 such parts, whereof AB, has but 10, and yet both of them to be in themselves of equal weight; it is certain, that whether the scales be empty, or whether in the scale D, we put 11 pound, and at E 10 pound, yet both of them shall equiponderate, because there is just such a disproportion in the length of the sides; AC, being unto AB, as 11 to 10 . . .

The manner how such deceitful ballances may be discovered, is by changing the weights into each other scale, and then the inequality will be manifest.

From the former grounds rightly apprehended, it is easie to conceive how a man may find out the just proportion of a weight, which in any point given, shall equiponderate to several weights given, hanging in several places of the Beam.

Some of these Ballances are made so exact, (those especially which the Refiners use) as to be sensibly turned with the eightieth part of a grain: which (though it may seem very strange) is nothing to what Capellus relates of one at Sedan, that would turn with the four hundredth part of a grain.

There are several contrivances to make use of these in measuring the weight of blows, the force of powder, the strength of strings, or other oblong substances, condensed air, the distinct proportion of several metals mixed together, the different gravity of divers bodies in the water, from what they have in the open air, with divers the like ingenious inquiries.

SOURCE: John Wilkins. *Mathematical Magick*, 4th ed (London, 1691), 14–17, 19–20

18 W. Molyneux
'DIOPTRICS'

Definitions.

I. Tab. 1. Fig. 1. ABCD is a Body of Glass, EFM is perpendicular to AB; suppose GF a Ray of Light falling inclined on the Glass AD, the Point F is called the Point of Incidence.

II. The Angle EFG comprehended between the Perpendicular and the Ray, is the Angle of Incidence (with Barrow, &c.) tho by many Dioptrick Writers 'tis called the Angle of Inclination; and its Complement AFG, is usually by them called the Angle of Incidence. But I shall use the Terms Inclination and Incidence promiscuously, always designing thereby the Angle comprehended between the Perpendicular EF and the Ray GF.

Experiments.

I. GF is a Ray of Light falling inclined on the Glass ABCD; this Ray coming out of a Rare Medium, as Air, into a Dense Medium, as Glass, does not proceed on its direct Course in a streight Line towards I; but at the Point of Incidence F 'tis bent or refracted towards the Perpendicular FM, and becomes the refracted Ray FH.

II. At H is its Point of Incidence again, from Glass a

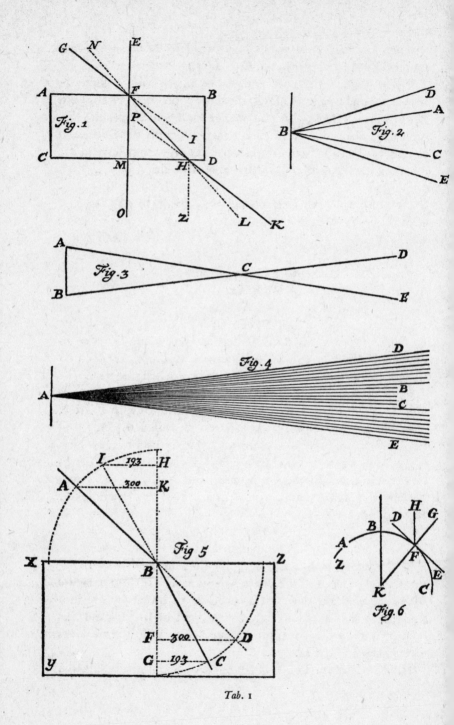

Tab. 1

Dense Medium to Air a Rare, and in this Passage, 'tis refracted from the Perpendicular HZ; so that instead of proceeding directly strait in FHL, 'tis refracted from the Perpendicular ZH, and becomes HK: So that if the Surface AB be parallel to the Surface DC, the Ray becomes again as if it had not been refracted; for now HK runs parallel to GF. The natural Reason of this Refraction is variously assigned by divers, and is properly of a Physical consideration; and what is offered therein, is little more than Hypothetical Conjecture. I shall not therefore mix Guesses with Demonstration: The Matter of Fact is manifest from Ten Thousand repeated Experiments, and this is sufficient to my purpose. But yet there is such an ingenious Hypothesis concerning this Matter, published in the *Acta Eruditorum Lipsia*, *Anno* 1682. *Mens. Junii*, pag. 185. by the Learned and Ingenious G. G. Leibnutzius, that I cannot omit inserting it in the Second Part of this Work, Chap. 1.

III. The Ray that falls perpendicular (as suppose EF a Ray) passes unrefracted, but all inclined Rays are refracted.

Definitions.

III. The Angle IFH, and KHL comprehended by the Ray directly prolonged, and the refracted Ray is the Angle of Refraction.

IV. The Angle HFM or ZHK comprised between the refracted Ray and Perpendicular is the refracted Angle.

V. Diverging Rays are those that spread and separate the farther from each other, as they flow farther from the Object. Tab. 1. Fig. 2. B is a radiating Point, AB, DB, CB, EB, are diverging Rays.

VI. Converging Rays are those that approach nigher each other, till they cross, and then become Diverging. Tab. 1. Fig. 3. A, B, are two Radiating Points in the Object AB, the Rays CA, CB, do Converge, till they cross in C, and then they become Diverging DC, EC.

VII. Parallel Rays are those that flowing from one and the same Point of a remote Object pass at the same distance, as to sense: But this is not to be strictly taken; for then the Rays

85

flowing from one and the same Point of an Object, cannot be parallel, for they always diverge: Yet when an Object is at such a great distance, and that parcel of Rays which is considered, is so small, that their Divergence is little or nothing considerable, these Rays are said to be parallel. Tab. 1. Fig. 4. A is a Point in an Object sending forth its diverging Rays AD, AB, AC, AE; let BC be the breadth of the Pupil or breadth of an Optick Glass: Here if the Point A be so remote, and BC be so small, that the Parcel of Rays BAC, do insensibly run parallel, then these are said to be parallel Rays.

VIII. Those radiating Points or Objects are said to be remote, whose distance from the Eye or Glass is so great, that the breadth of the Pupil or Glass, in respect thereto, is inconsiderable.

IX. Those radiating Points or Objects are said to be nigh, when there is a sensible proportion between the Pupils or Glasses breadth, and the distance; so that the Rays flowing from any single Point thereof do not run parallel to each other, but diverge considerably in respect of the Pupils or Glasses breadth.

Experiments.

IV. Tab. 1. Fig. 5. ZXY is a Body of Glass, HBG is perpendicular to ZX, AB is a Ray falling on this Glass, and the Angle of Inclination or Incidence is HBA = DBG. Kepler tells us, that under 30 deg. of Inclination from Air to Glass, the Angle of Refraction DBC is $\frac{1}{3}$ of the Inclination, therefore the refracted Angle CBG is $\frac{2}{3}$ of the Inclination. Wherefore we lay down the following Proportions, as confirmed by Kepler's Experiments, and usually retain'd by most Optick Writers.

L Inclination DBG : refracted L CBG : : 3 : 2
L Inclination DBG : L Refraction DBC : : 3 : 1
L Refraction DBC : refracted L CBG : : 1 : 2

V. But suppose the Ray CB to proceed from Glass to Air; at B 'tis refracted from the Perpendicular BH, and becomes BA; here the Inclination is CBG = HBI, and then

86

L Incidence HBI : refracted *L* HBA : : 2 : 3
L Inclination HBI : *L* Refraction IBA : : 2 : 1
L Refraction IBA : refracted *L* HBA : : 1 : 3

These Propositions in the 4th. and 5th. Experiments we shall retain in the following Demonstrations, for the Ease and Plainness thereof. But in Calculation we shall observe the Proportion that follows in the 6th. Experiment.

SOURCE: W. Molyneux. *Dioptrica Nova* (London, 1692), 1–4 and Table 1

CHEMISTRY

Two main methods were employed in the seventeenth century to advance men's understanding of chemistry: alchemical and mechanical. The alchemical method, inheriting a long tradition of rhetoric and symbols, remained most popular. It was practised by the continentals Van Helmont and Paracelsus, in the early part of the seventeenth century, and in England by thinkers of the stature of Boyle and Newton. Men sought to understand the transmutations of substances, and sought the Elixir of Life, which they thought would provide the answer to the whole arcana of nature's secrets. They hoped, through alchemical jargon, to understand the processes at work behind the universe, which would enable them to control the universe and to control their own lives.

Those who sought to apply the mechanical method to chemistry did so through experiments involving for example the weighing of plants and the soil in which they grew to gauge the chemicals used up in the growth process. Sanctorius weighed human beings before and after meals and attempted to gauge the quantities of food and other intakes absorbed and utilised by the body. They hoped that quantifying and mathematics would produce the same results in chemistry which they had produced with such success in many other disciplines.

In the extract below from Boyle's Sceptical Chymist *(19) Carneades criticises extant chemical theory, whether it be the Peripatetic or the hermetic systems. He denies that matter can be resolved by fire into three simple elements, or into five main ingredients. Instead of any rigid theory based upon earth, fire, water, phlegm, oil, sulphur, mercury and spirit, Carneades pleads for an experimental chemistry using any*

technique available to break down compounds and mixtures and analyse
them further. He confesses at the end that his own scepticism fails to
satisfy him, but that he cannot accept the doctrines of either Aristotle or
Paracelsus

19 Robert Boyle
'THE SCEPTICAL CHYMIST'

THE CONCLUSION

THESE last words of Carneades being soon after followed by a
noise which seemed to come from the place where the rest of
the company was, he took it for a warning, that it was time for
him to conclude or break off his discourse; and told his friend;
By this time I hope you see, Eleutherius, that if Helmont's
experiments be true, it is no absurdity to question whether that
doctrine be one, that doth not assert any elements in the sence
before explained. But because that, as divers of my arguments
suppose the marvellous power of the alkahest in the analyzing
of bodies, so the effects ascribed to that power are so unparalleled
and stupendous, that though I am not sure but that there may
be such an agent, yet little less than $\alpha\nu\tau o\Psi\acute{\iota}\alpha$ seems requisite to
make a man sure there is. And consequently I leave it to you to
judge, how farr those of my arguments that are built upon
alkahestical operations are weakned by that liquors being
matchless; and shall therefore desire you not to think that I
propose this paradox that rejects all elements, as an opinion
equally probable with the former part of my discourse. For by
that, I hope, you are satisfied, that the arguments, wont to be
brought by chymists to prove that all bodies consist of either
three principles, or five, are far from being so strong as those
that I have employed to prove, that there is not any certain and
determinate number of such principles or elements to be met
with universally in all mixt bodies. And I suppose I need not
tell you, that these anti-chymical paradoxes might have been
managed more to their advantage; but that having not confined my curiosity to chymical experiments, I, who am but a
young man, and younger chymist, can yet be but slenderly

furnished with them, in reference to so great and difficult a task as you imposed upon me: besides that, to tell you the truth, I durst not employ some even of the best experiments I am acquainted with, because I must not yet disclose them; but, however, I think I may presume that what I have hitherto discoursed will induce you to think, that chymists have been much more happy in finding experiments than the causes of them; or in assigning the principles by which they may best be explained. And indeed, when in the writing of Paracelsus I meet with such phantastick and unintelligible discourses as that writer often puzzels and tires his reader with, fathered upon such excellent experiments, as though he seldom clearly teaches, I often find he knew; methinks the chymists, in their searches after truth, are not unlike the navigators of Solomon's Tarshish fleet, who brought home from their long and tedious voyages, not only gold, and silver, and ivory, but apes and peacocks too; for so the writings of several (for I say not, all) of your hermetick philosophers present us, together with divers substantial and noble experiments, theories, which either like peacocks' feathers make a great shew, but are neither solid nor useful; or else like apes, if they have some appearance of being rational, are blemished with some absurdity or other, that when they are attentively considered, make them appear ridiculous.

Carneades having thus finished his discourse against the received doctrines of the elements, Eleutherius judging he should not have time to say much to him before their separation, made some haste to tell him; I confess, Carneades, that you have said more in favour of your paradoxes than I expected. For though divers of the experiments you have mentioned are no secrets, and were not unknown to me, yet besides that you have added many of your own unto them, you have laid them together in such a way, and applyed them to such purposes, and made such deductions from them, as I have not hitherto met with.

But though I be therefore inclined to think, that Philoponus, had he heard you, would scarce have been able in all points to defend the chymical hypothesis against the arguments wherewith you have opposed it; yet methinks that however your objections seem to evince a great part of what they pretend to, yet they evince it not all; and the numerous tryals of those you call the vulgar chymists, may be allowed to prove something too.

Wherefore, if it be granted you that you have made it probable.

First, that the differing substances into which mixt bodies are wont to be resolved, by the fire are not of a pure and an elementary nature, especially for this reason, that they yet retain so much of the nature of the concrete that afforded them, as to appear to be yet somewhat compounded, and oftentimes to differ in one concrete from principles of the same denomination in another:

Next, that as to the number of these differing substances, neither is it precisely three, because in most vegetable and animal bodies earth and phlegme are also to be found among their ingredients; nor is there any one determinate number into which the fire (as it is wont to be employed) does precisely and universally resolve all compound bodies whatsoever, as well minerals as others that are reputed perfectly mixt.

Lastly, that there are divers qualities which cannot well be referred to any of these substances, as if they primarily resided in it and belonged to it; and some other qualities, which though they seem to have their chief and most ordinary residence in some one of these principles or elements of mixt bodies, are not yet so deducible from it, but that also some more general principles must be taken in to explicate them.

If, I say, the chymists (continues Eleutherius) be so liberall as to make you these three concessions, I hope you will, on your part, be so civil and equitable as to grant them these three other propositions, namely;

First, that divers mineral bodies, and therefore probably all the rest, may be resolved into a saline, a sulphureous, and a mercurial part; and that almost all vegetable and animal concretes may, if not by the fire alone, yet by a skilfull artist employing the fire as his chief instrument, be divided into five differing substances, salt, spirit, oyle, phlegme and earth; of which the three former by reason of their being so much more operative than the two latter, deserve to be lookt upon as the three active principles, and by way of eminence to be called the three principles of mixt bodies.

Next, that these principles, though they be not perfectly devoid of all mixture, yet may without inconvenience be stiled the elements of compounded bodies, and bear the names of those substances which they most resemble, and which are manifestly predominant in them; and that especially for this reason, that none of these elements is divisible by the fire into four or five differing substances, like the concrete whence it was separated.

Lastly, that divers of the qualities of a mixt body, and especially the medical virtues, do for the most part lodge in some one or other of its principles, and may therefore usefully be sought for in that principle severed from the others.

And in this also (pursues Eleutherius) methinks both you and the chymists may easily agree, that the surest way is to learn by particular experiments, what differing parts particular bodies do consist of, and by what wayes (either actual or potential fire) they may best and most conveniently be separated, as without relying too much upon the fire alone, for the resolving of bodies, so without fruitlessly contending to force them into more elements than nature made them up of, or strip the severed principles so naked, as by making them exquisitely elementary to make them almost useless.

These things (subjoynes Eleu.) I propose, without despairing to see them granted by you; not only because I know that you so much prefer the reputation of candour before that of subtility, that your having once supposed a truth would not hinder you from imbracing it when clearly made out to you; but because, upon the present occasion, it will be no disparagement to you to recede from some of your paradoxes, since the nature and occasion of your past discourse did not oblige you to declare your own opinions, but only to personate an antagonist of the chymists. So that (concludes he, with a smile) you may now by granting what I propose, add the reputation of loving the truth sincerely to that of having been able to oppose it subtilly.

Carneades's haste forbidding him to answer this crafty piece of flattery; Till I shall (saies he) have an opportunity to acquaint you with my own opinions about the controversies I have been discoursing of, you will not I hope, expect I should declare my own sense of the argument I have employed. Wherefore I shall only tell you thus much at present; that though not only an acute naturalist, but even I myself could take plausible exceptions at some of them; yet divers of them too are such as will not perhaps be readily answered, and will reduce my adversaries, at least, to alter and reform their hypothesis, I perceive I need not mind you that the objections I made against the quaternary of elements and ternary of principles needed not to be opposed so much against the doctrines themselves, (either of which, especially the latter, may be much more probably maintained than hitherto it seems to have been, by those writers for it I have met with) as against the unaccurateness and the unconcludingness of the analytical experiments vulgarly relyed on to demonstrate them.

And therefore, if either of the two examined opinions, or any other theory of elements, shall upon rational and experimental grounds be clearly made out to me; 'tis obliging, but not irrational, in you to expect, that I shall not be so farr in love with my disquieting doubts, as not to be content to change them for undoubted truths. And (concludes Carneades smiling) it were no great disparagement for a sceptick to confesse to you, that as unsatisfyed as the past discourse may have made you think me with the doctrines of the Peripateticks, and the chymists, about the elements and principles, I can yet so little discover what to acquiesce in, that perchance the enquiries of others have scarce been more unsatisfactory to me, than my own have been to myself.

SOURCE: Robert Boyle. *The Sceptical Chymist* (reprinted 1967), 226–30

MAGNETISM

It was William Gilbert who at the opening of the seventeenth century thought magnetism essential to a complete understanding of the world's movements, geography, and to understanding the whole of natural philosophy. It was Gilbert, English classicist, man of letters, and physician, who provided the first full-length treatment of the magnet, combining experiments with simple loadstones with commentary and criticism of the Ancients and their views on the subject.

Those who followed after Gilbert were not convinced that in magnetism lay a solution to the motion of the heavens. Galileo criticised Gilbert and his argument that a freely suspended loadstone would, like the earth acting as a great magnet, freely revolve around its notional centre (20). Galileo suggested that a magnet is not a fair analogy for the action and motion of the earth.

Gilbert's enquiries helped to base the physics of magnets more soundly on empirical foundations. Later in the century the interest in the force of magnetism declined as gravity increased in importance as a way of explaining the movements of the planets. Magnetism remained crucial to the development of the compass, and became a study taken up by the practitioners and navigators keen to understand the science behind the technology they were attempting to develop. The seventeenth century looked to Gilbert for the most complete and provocative explanation of the force of magnetism; it was he who pioneered work on magnetic variation and the dip of compass needles. The first part of the extract from Gilbert's De Magnete *(published in 1600) illustrates the way in which he built up a science of magnetism from basic principles and simple observations; the second shows the way he saw a parallel between the operation of a loadstone and the motion of the earth.*

20 William Gilbert
'MAGNETISM AND THE WORLD PICTURE'

ONE LOADSTONE APPEARS TO ATTRACT ANOTHER IN THE NATURAL
POSITION; BUT IN THE OPPOSITE POSITION REPELS IT AND BRINGS
IT TO RIGHTS

FIRST we have to describe in popular language the potent and
familiar properties of the stone; afterward, very many subtile
properties, as yet recondite and unknown, being involved in
obscurities, are to be unfolded; and the causes of all these
(nature's secrets being unlocked) are in their place to be
demonstrated in fitting words and with the aid of apparatus.
The fact is trite and familiar, that the loadstone attracts iron; in
the same way, too, one loadstone attracts another. Take the
stone on which you have designated the poles, N. and S., and
put it in its vessel so that it may float; let the poles lie just in the
plane of the horizon, or at least in a plane not very oblique to it;
take in your hand another stone the poles of which are also
known, and hold it so that its south pole shall lie toward the
north pole of the floating stone, and near it alongside; the floating loadstone will straightway follow the other (provided it be
within the range and dominion of its powers), nor does it cease
to move nor does it quit the other till it clings to it, unless, by
moving your hand away, you manage skilfully to prevent the
conjunction. In like manner, if you oppose the north pole of the
stone in your hand to the south pole of the floating one, they
come together and follow each other. For opposite poles attract
opposite poles. But, now, if in the same way you present N. to
N. or S. to S., one stone repels the other; and as though a helms-
man were bearing on the rudder it is off like a vessel making all
sail, nor stands nor stays as long as the other stone pursues. One
stone also will range the other, turn the other around, bring it
to right about and make it come to agreement with itself. But
when the two come together and are conjoined in nature's
order, they cohere firmly. For example, if you present the north

94

pole of the stone in your hand to the Tropic of Capricorn (for so we may distinguish with mathematical circles the round stone or terrella, just as we do the globe itself) or to any point between the equator and the south pole: immediately the floating stone turns round and so places itself that its south pole touches the north pole of the other and is most closely joined to it. In the same way you will get like effect at the other side of the equator by presenting pole to pole; and thus by art and contrivance we exhibit attraction and repulsion, and motion in a circle toward the concordant position, and the same movements to avoid hostile meetings. Furthermore, in one same stone we are thus able to demonstrate all this: but also we are able to show how the self-same part of one stone may by division become either north or south. Take the oblong stone *ad* in which *a* is the north pole and *d* the south. Cut the stone in two equal parts, and put part *a* in a vessel and let it float in water.

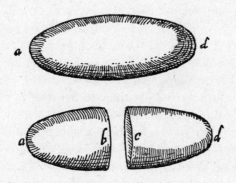

You will find that *a*, the north point, will turn to the south as before; and in like manner the point *d* will move to the north, in the divided stone, as before division. But *b* and *c*, before connected, now separated from each other, are not what they were before. *b* is now south while *c* is north. *b* attracts *c*, longing for union and for restoration of the original continuity. They are two stones made out of one, and on that account the *c* of one turning toward the *b* of the other, they are mutually attracted, and, being freed from all impediments and from their own weight, borne as they are on the surface of the water, they come together and into conjunction. But if you bring the part or point *a* up to *c* of the other, they repel one another and turn away; for by such a position of the parts nature is crossed and the form of the stone is perverted: but nature observes strictly the laws it has imposed upon bodies: hence the flight of one part from the undue position of the other, and hence the discord unless everything is arranged exactly according to nature. And nature will not suffer an unjust and inequitable peace, or an unjust and inequitable peace and agreement, but makes war and employs force to make bodies acquiesce fairly and justly. Hence, when rightly arranged, the parts attract each other, i.e., both stones, the weaker and the stronger, come together and with all their might tend to union: a fact manifest in all loadstones, and not, as Pliny supposed, only in those from Ethiopia. The Ethiopic stones if strong, and those brought from China, which are all powerful stones, show the effect most quickly and most plainly, attract with most force in the parts nighest the pole, and keep turning till pole looks straight on pole. The pole of a stone has strongest attraction for that part of another stone which answers to it (the *adverse* as it is called); e.g., the north pole of one has strongest attraction for, has the most vigorous pull on, the south part of another: so too it attracts iron more powerfully, and iron clings to it more firmly, whether previously magnetized or not. Thus it has been settled by nature, not without reason, that the parts nigher the pole shall have the greatest attractive force; and that in the pole itself shall be the seat, the throne as it were, of a high and splendid power; and that magnetic bodies brought near thereto shall be attracted most powerfully and relinquished with most reluctance. So, too, the poles are readiest to spurn and drive away what is presented to them amiss, and what is inconformable and foreign.

THAT THE TERRESTRIAL GLOBE IS MAGNETIC AND IS A LOADSTONE;
AND JUST AS IN OUR HANDS THE LOADSTONE POSSESSES ALL THE
PRIMARY POWERS (FORCES) OF THE EARTH, SO THE EARTH BY
REASON OF THE SAME POTENCIES LIES EVER IN THE SAME DIREC-
TION IN THE UNIVERSE

BEFORE we expound the causes of the magnetic movements
and bring forward our demonstrations and experiments touching matters that for so many ages have lain hid – the real
foundations of terrestial philosophy – we must formulate our
new and till now unheard-of view of the earth, and submit it to
the judgment of scholars. When it shall have been supported
with a few arguments of *prima facie* cogency, and these shall
have been confirmed by subsequent experiments and demonstrations, it will stand as firm as aught that ever was proposed
in philosophy, backed by ingenious argumentation, or but-
tressed by mathematical demonstrations. The terrestrial mass
which together with the world of waters produces the spherical
figure and our globe, inasmuch as it consists of firm durable
matter, is not easily altered, does not wander nor fluctuate with
indeterminate movements like the seas and the flowing streams;
but in certain hollows, within certain bounds, and in many
veins and arteries, as it were, holds the entire volume of liquid
matter, nor suffers it to spread abroad and be dissipated. But
the solid mass of the earth has the greater volume and holds pre-
eminence in the constitution of our globe. Yet the water is
associated with it, though only as something supplementary and
as a flux emanating from it; and from the beginning it is inti-
mately mixed with the smallest particles of earth and is innate
in its substance. The earth growing hot emits it as vapor, which
is of the greatest service to the generation of things. But the
strong foundation of the globe, its great mass, is that terrene
body, far surpassing in quantity the whole aggregate of fluids
and waters whether in combination with earth or free (whatever vulgar philosophers may dream about the magnitudes and
proportions of their elements); and this mass makes up most of

the globe, constituting nearly its whole interior framework, and of itself taking on the spherical form. For the seas do but fill certain not very deep hollows, having very rarely a depth of a mile, and often not exceeding 100 or 50 fathoms. This appears from the observations of navigators who have with line and sinker explored their bottoms. In view of the earth's dimension, such depressions cannot much impair the spheroidal shape of the globe. Still the portion of the earth that ever comes into view for man or that is brought to the surface seems small indeed, for we cannot penetrate deep into its bowels, beyond the *débris* of its outermost efflorescence, hindered either by the waters that flow as through veins into great mines; or by the lack of wholesome air necessary to support the life of the miners; or by the enormous cost of executing such vast undertakings, and the many difficulties attending the work. Thus we cannot reach the inner parts of the globe, and if one goes down, as in a few mines, 400 fathoms, or (a very rare thing) 500 fathoms, it is something to make every one wonder. But how small, how almost null, is the proportion of 500 fathoms to the earth's diameter, 6,872 miles, can be easily understood. So we do only see portions of the earth's circumference, of its prominences; and everywhere these are either loamy, or argillaceous, or sandy; or consist of organic soils or marls; or it is all stones and gravel; or we find rock-salt, or ores, or sundry other metallic substances. In the depths of the ocean and other waters are found by mariners, when they take soundings, ledges and great reefs, or bowlders, or sands, or ooze. The Aristotelian element, earth, nowhere is seen, and the Peripateties are misled by their vain dreams about elements. But the great bulk of the globe beneath the surface and its inmost parts do not consist of such matters; for these things had not been were it not that the surface was in contact with and exposed to the atmosphere, the waters, and the radiations and influences of the heavenly bodies; for by the action of these are they generated and made to assume many different forms of things, and to change perpetually. Still do they imitate the inner parts and resemble their source, because their matter is of the earth, albeit they have lost the prime qualities and the true nature of terrene matter; and they bear toward the earth's centre and cohere to the globe and cannot be parted from it save by force. Yet the loadstone and all magnetic bodies — not only the stone but all magnetic, homogenic matter — seem to contain within themselves the potency of the earth's core and of its inmost viscera, and to have and comprise whatever in the earth's substance is privy and inward: the loadstone possesses the actions peculiar to the globe, of attraction, polarity, revolution, of taking position in the universe according to the law of the whole; it contains the supreme excellencies of the globe and orders them: all this is token and proof of a certain eminent combination and of a most accordant nature. For, if among bodies one sees aught that moves and breathes and has senses and is governed and impelled by reason, will he not, knowing and seeing this, say that here is a man or something more like man than a stone or a stalk? The loadstone far surpasses all other bodies around us in the virtues and properties that pertain to the common mother of all; but those properties have been very little understood and noted by philosophers. Toward it, as we see in the case of the earth, magnetic bodies tend from all sides, and adhere to it; it has poles — not mathematical points, but natural points of force that through the co-operation of all its parts excel in prime efficiency; such poles exist also in the same way in the globe, and our forefathers always sought them in the heavens. Like the earth, it has an equator, a natural line of demarkation between the two poles; for of all the lines drawn by mathematicians on the terrestrial globe, the equator (as later will appear) is a natural boundary, and not merely a mathematical circle. Like the earth, the loadstone has the power of direction and of standing still at north and south; it has also a circular motion to the earth's position, whereby it adjusts itself to the earth's law. It follows the elevations and depressions of the earth's poles, and conforms precisely to them: according to the position of the earth and of the locality, it naturally and of itself elevates its poles above the horizon, or depresses them. The loadstone derives properties from the earth *ex tempore*, and acquires verticity; and iron is affected by the verticity of the globe as it is affected by a load-stone. Magnetic bodies are governed and regulated by the earth, and they are subject to the earth in all their movements. All the movements of the loadstone are in accord with the geometry and form of the earth and are strictly controlled thereby, as will later be proved by conclusive experiments and diagrams; and the greater part of the visible earth is also magnetic, and has magnetic movements, though it is defaced by all sorts of waste matter and by no end of transformations. Why, then, do we not recognize this primary and homogeneous earth-substance, likest of all substances to the inmost nature, to the very marrow, of the earth itself, and nearest to it? For not any of the other mixed earths – those suitable for agriculture, – not any of the metalliferous veins, no stones, no sands, no other fragments of the globe that come under our notice, possess such stable, such distinctive virtues. Yet we do not hold the whole interior of this our globe to be of rock or of iron, albeit the learned Franciscus Maurolycus* deems the earth in its interior to consist throughout of rigid rock. For not every loadstone that we find is a stone, being sometimes like a clod of earth, or like clay, or like iron; consisting of various materials compacted into hardness, or soft, or by heat reduced to the metallic state; and in the earth's surface formations, according to circumstances of place, of the bodies around it, and of its matrix in the mine, a magnetic substance is distinguished by divers qualities and by adventitious accretions, as we see in marl, in some stones, and in iron ores. But the true earth-matter we hold to be a solid body homogeneous with the globe, firmly coherent, endowed with a primordial and (as in the other globes of the universe) an energic form. By being so fashioned, the earth has a fixed verticity, and necessarily revolves with an innate whirling

* Francis Maurolico—Maurolycus, Marullo (1494–1575)—was abbot of Messina, where he publicly taught mathematics, and was quite a voluminous writer upon different scientific subjects, his works including very able treatises, more particularly on the sphere, on astronomical instruments, etc. A full account of his life and writings was issued at Messina (Messanæ) in 1613, the date and place likewise of his very interesting magnetical book entitled 'Problemata mechanica, cum appendice et ad magnetum, et ad pixidem nauticam pertinentia.'

motion: this motion the loadstone alone of all the bodies around us possesses genuine and true, less spoilt by outside interferences, less marred than in other bodies, – as though the motion were an homogenous part taken from the very essence of our globe. This pure native iron is produced when homogenic portions of the earth's substance coalesce to form a metallic vein; loadstone is produced when they are transformed into metallic stone or a vein of the finest iron or steel; so, too, rather imperfect homogenic material collects to form other iron ores – just as many parts of the earth, even parts that rise above the general circumference, are of homogenic matter, only still more debased. Native iron is iron fused and reduced from homogenic matters, and coheres to earth more tenaciously than the ores themselves. Such, then, we consider the earth to be in its interior parts; it possesses a magnetic homogenic nature. On this more perfect material (foundation) the whole world of things terrestrial, which, when we search diligently, manifests itself to us everywhere, in all the magnetic metals and iron ores and marls, and multitudinous earths and stones; but Aristotle's 'simple element', and that most vain terrestrial phantasm of the Peripatetics,– formless, inert, cold, dry, simple matter, the substratum of all things, having no activity, – never appeared to any one even in dreams, and if it did appear would be of no effect in nature. Our philosophers dreamt only of an inert and simple matter. Cardan thinks the loadstone is not a stone of any species, but that it is, as it were, a perfect portion of a certain kind of earth that is absolute, whereof a proof is its abundance, for there is no place where it is not found. He says that this kind of conceptive, generative earth, possessed of an affinity like that of the marriage tie, is perfected when it has been placed in contact with, or received the fecundating influence of, the masculine or Herculean stone, it having been, moreover, shown in a previous proposition (*Libro de Proportionibus*) that the loadstone is true earth.*

A strong loadstone shows itself to be of the inmost earth, and

---

* Consult Cardan's Works, Lugduni (Lyons), 1663 ed., Vol. II, *De Exemplo*. . . ., pages 539, 546, Vol. III, Lib. V, Cap. XVII–XIX, Vol. X, Cap. VI, page 12.

in innumerable experiments proves its claim to the honor of possessing the primal form of things terrestrial, in virtue of which the earth itself remains in its position and is directed in its movements. So a weak loadstone, and all iron ore, all marls and argillaceous and other earths (some more, some less, according to the difference of their humors and the varying degrees in which they have been spoilt by decay), retain, deformed, in a state of degeneration from the primordial form, magnetic properties, powers, that are conspicuous and in the true sense telluric. For not only does metallic iron turn to the poles, not only is one loadstone attracted by another and made to revolve magnetically, but so do (if prepared) all iron ores and even other stones, as slates from the Rhineland, the black slates (*ardoises*, as the French call them) from Anjou, which are used for shingles, and other sorts of fissile stone of different colors; also clays, gravel, and several sorts of rock; and, in short, all of the harder earths found everywhere, provided only they be not fouled by oozy and dank defilements like mud, mire, heaps of putrid matter, or by the decaying remains of a mixture of organic matters, so that a greasy slime oozes from them, as from marl, – they are all attracted by the loadstone, after being prepared simply by the action of fire and freed from their excrementitious humor; and as by the loadstone, so, too, are they magnetically attracted and made to point to the poles by the earth itself, therein differing from all other bodies; and by this innate force they are made to conform to the ordering and planning of the universe and the earth, as later will appear. Thus every separate fragment of the earth exhibits in indubitable experiments the whole impetus of magnetic matter; in its various movements it follows the terrestrial globe and the common principle of motion.

SOURCE: William Gilbert. *De Magnete*, translated by P. F. Mottelay (London 1893), 28–31, 64–71.

ANATOMY AND MEDICINE

The quest for knowledge of nature's ways and means was never more strenuous than when it was concerned with knowledge of the functioning of the human body. Life was short, the threat of death ever present; men sought after security for their lives by devoting their energies to cures and palliatives, to anatomy and medicine. When the century began, much of the Galenist fabric remained and Aristotle, as in all branches of learning, was an undisputed master of the craft. Given the highly technical and professional education that men had to undergo to obtain their medical qualifications – unlike any other science which would as often as not form no part of any recognised curriculum of a university – the Ancients held sway far longer, as they were the guarantors of academic rigour and respectability. But here as elsewhere criticism and empiricism soon made more incursions into the ancient pattern of ideas. It was discovered that Aristotle had perhaps never seen the internal workings of a human body and that his Anatomy was based upon analysis of apes. New lecture theatres like those constructed at Padua allowed students to observe a practical at first hand. Anatomists discovered discrepancies in the highly prized text of Galen; while a new comparative anatomy was developed in works like that of Tyson's, which compared the anatomy of the pygmy and the orang-utan.

At Padua and Leyden, and to a lesser extent at Oxford and Cambridge, anatomy was pursued with vigour. Harvey reappraised many of the Aristotelian and Galenical ideas and is remembered chiefly for his work on the circulation of the blood. The crucial passages relating to this are reproduced below (21). It is interesting to compare them with Gassendi's work illustrating that he was satisfied with the preceding theory that blood travelled from right to left in the heart through a passage in the septum (22). The passage also shows that Gassendi was in touch with Fludd, the English iatrochemist and pansophist. Harvey's opening gives a prelude on the nature of anatomy and shows his deference to and criticism of Galen and Aristotle: they are crucial authorities, he suggests, but often incorrect ones.

Van Helmont was an important European chemist, and his views are canvassed here on the nature of Paracelsus and his message (23). Van Helmont was a critical Paracelsian, who believed in the application of chemistry to medicine, and who with Paracelsus believed in the connections between the world of magic and the world of the physician. This tradition of thought gave rise to the chemical remedies hawked about Europe from the mid-seventeenth century onwards.

Digby's piece on weapon salves (24A) illustrates one of the most popular of beliefs from a most popular book; he shows how he believes in the magnetic cure of wounds. There were many similar theories in common circulation, among the intellectuals as well as rural widows and herb women (24B). The section is completed by Malpighi's work on the lungs (25), the work of an important microscopist who applied his skills with optics to the business of studying the human anatomy.

21 William Harvey
'CIRCULATION OF THE BLOOD'

OF THE MOTION, ACTION, AND OFFICE OF THE HEART

From these and other observations of the like kind, I am persuaded it will be found that the motion of the heart is as follows:

First of all, the auricle contracts, and in the course of its contraction throws the blood (which it contains in ample quantity as the head of the veins, the store-house and cistern of the blood) into the ventricle, which being filled, the heart raises itself straightway, makes all its fibres tense, contracts the ventricles, and performs a beat, by which beat it immediately sends the blood supplied to it by the auricle into the arteries; the right ventricle sending its charge into the lungs by the vessel which is called vena arteriosa, but which, in structure and function, and all things else, is an artery; the left ventricle sending its charge into the aorta, and through this by the arteries to the body at large.

These two motions, one of the ventricles, another of the auricles, take place consecutively, but in such a manner that there is a kind of harmony or rhythm preserved between them, the two concurring in such wise that but one motion is apparent, especially in the warmer blooded animals, in which the movements in question are rapid. Nor is this for any other reason than it is in a piece of machinery, in which, though one wheel gives motion to another, yet all the wheels seem to move simultaneously; or in that mechanical contrivance which is adapted to firearms, where the trigger being touched, down comes the flint, strikes against the steel, elicits a spark, which falling among the powder, is ignited, upon which the flame extends, enters the barrel, causes the explosion, propels the ball and the mark is attained – all of which incidents, by reason of the celerity with which they happen, seem to take place in the twinkling of an eye. So also in deglutition: by the elevation of the root of the tongue, and the compression of the mouth, the food or drink is pushed into the fauces, the larynx is closed by its own muscles, and the epiglottis, whilst the pharynx, raised and opened by its muscles no otherwise than is a sac that is to be filled, is lifted up, and its mouth dilated; upon which, the mouthful being received, it is forced downwards by the transverse muscles, and then carried farther by the longitudinal ones. Yet are all these motions, though executed by different and distinct organs, performed harmoniously, and in such order, that they seem to constitute but a single motion and act, which we call deglutition.

Even so does it come to pass with the motions and action of the heart, which constitute a kind of deglutition, a transfusion of the blood from the veins to the arteries. And if any one, bearing these things in mind, will carefully watch the motions of the heart in the body of a living animal, he will perceive not only all the particulars I have mentioned, viz., the heart becoming erect, and making one continuous motion with its auricles; but farther, a certain obscure undulation and lateral inclination in the direction of the axis of the right ventricle [the organ], twisting itself slightly in performing its work. And indeed every one may see, when a horse drinks, that the water is drawn in and transmitted to the stomach at each movement of the throat, the motion being accompanied with a sound, and yielding a pulse both to the ear and the touch; in the same way it is with each motion of the heart, when there is the delivery of a quantity of blood from the veins to the arteries, that a pulse takes place, and can be heard within the chest.

The motion of the heart, then, is entirely of this description, and the one action of the heart is the transmission of the blood and its distribution, by means of the arteries, to the very extremities of the body; so that the pulse which we feel in the arteries is nothing more than the impulse of the blood derived from the heart.

Whether or not the heart, besides propelling the blood, giving it motion locally, and distributing it to the body, adds anything else to it, – heat, spirit, perfection, – must be inquired into by and by, and decided upon other grounds. So much may suffice at this time, when it is shown that by the action of the heart the blood is transfused through the ventricles from the veins to the arteries, and distributed by them to all parts of the body.

So much, indeed, is admitted by all [physiologists], both from the structure of the heart and the arrangement and action of its valves. But still they are like persons purblind or groping about in the dark; and then they give utterance to diverse, contradictory, and incoherent sentiments, delivering many things upon conjecture, as we have already had occasion to remark.

The grand cause of hesitation and error in this subject appears to me to have been the intimate connexion between the heart and the lungs. When men saw both the vena arteriosa [or pulmonary artery] and the arteriae venosae [or pulmonary veins] losing themselves in the lungs, of course it became a puzzle to them to know how or by what means the right ventricle should distribute the blood to the body, or the left draw it from the venae cavae. This fact is borne witness to by Galen, whose words, when writing against Erasistratus in regard to the origin and use of the veins and the coction of the blood, are the following: 'You will reply,' he says, 'that the effect is so; that the blood is prepared in the liver, and is thence transferred to the heart to receive its proper form and last perfection; a statement which does not appear devoid of reason; for no great and perfect work is ever accomplished at a single effort, or receives its final polish from one instrument. But if this be actually so, then show us another vessel which draws the absolutely perfect blood from the heart, and distributes it as the arteries do the spirits over the whole body.' Here then is a reasonable opinion not allowed, because, forsooth, besides not seeing the true means of transit, he could not discover the vessel which should transmit the blood from the heart to the body at large!

But had any one been there in behalf of Erasistratus, and of that opinion which we now espouse, and which Galen himself acknowledges in other respects consonant with reason, to have pointed to the aorta as the vessel which distributes the blood from the heart to the rest of the body, I wonder what would have been the answer of that most ingenious and learned man? Had he said that the artery transmits spirits and not blood, he would indeed sufficiently have answered Erasistratus, who imagined that the arteries contained nothing but spirits; but then he would have contradicted himself, and given a foul denial to that for which he had keenly contended in his writings against this very Erasistratus, to wit, that blood in substance is contained in the arteries, and not spirits; a fact which he demonstrated not only by many powerful arguments, but by experiments.

But if the divine Galen will here allow, as in other places he does, 'that all the arteries of the body arise from the great artery, and that this takes its origin from the heart; that all these vessels naturally contain and carry blood; that the three semi-lunar valves situated at the orifice of the aorta prevent the return of the blood into the heart, and that nature never connected them with this, the most noble viscus of the body, unless for some most important end'; if, I say, this father of physic admits all these things, – and I quote his own words, – I do not see how he can deny that the great artery is the very vessel to carry the blood, when it has attained its highest term of perfection, from the heart for distribution to all parts of the body. Or would he perchance still hesitate, like all who have come after him, even to the present hour, because he did not perceive the route by which the blood was transferred from the veins to the arteries, in consequence, as I have already said, of the intimate

107

connexion between the heart and the lungs? And that this difficulty puzzled anatomists not a little, when in their dissections they found the pulmonary artery and left ventricle full of thick, black, and clotted blood, plainly appears, when they felt themselves compelled to affirm that the blood made its way from the right to the left ventricle by sweating through the septum of the heart. But this fancy I have already refuted. A new pathway for the blood must therefore be prepared and thrown open, and being once exposed, no further difficulty will, I believe, be experienced by anyone in admitting what I have already proposed in regard to the pulse of the heart and arteries, viz. the passage of the blood from the veins to the arteries, and its distribution to the whole of the body by means of these vessels.

OF THE QUANTITY OF BLOOD PASSING THROUGH THE HEART FROM THE VEINS TO THE ARTERIES; AND OF THE CIRCULAR MOTION OF THE BLOOD

Thus far I have spoken of the passage of the blood from the veins into the arteries, and of the manner in which it is transmitted and distributed by the action of the heart; points to which some, moved either by the authority of Galen or Columbus, or the reasonings of others, will give in their adhesion. But what remains to be said upon the quantity and source of the blood which thus passes, is of so novel and unheard of character, that I not only fear injury to myself from the envy of a few, but I tremble lest I have mankind at large for my enemies, so much doth wont and custom, that become as another nature, and doctrine once sown and that hath struck deep root, and respect for antiquity influence all men. Still the die is cast, and my trust is in my love of truth, and the candour that inheres in cultivated minds. And sooth to say, when I surveyed my mass of evidence, whether derived from vivisections, and my various reflections on them, or from the ventricles of the heart, and the vessels that enter into and issue from them, the symmetry and size of these conduits, – for nature doing nothing in vain, would never have given them so large a relative size without a purpose, – or from the arrangement and intimate structure of the valves in particular, and of the other parts of the heart in general, with many things besides, I frequently and seriously bethought me, and long revolved in my mind, what might be the quantity of blood which was transmitted, in how short a time its passage might be effected, and the like; and not finding it possible that this could be supplied by the juices of the ingested aliment without the veins on the one hand becoming drained, and the arteries on the other getting ruptured through the excessive charge of blood, unless the blood should somehow find its way from the arteries into the veins, and so return to the right side of the heart; I began to think whether there might not be A MOTION, AS IT WERE, IN A CIRCLE. Now this I afterwards found to be true; and I finally saw that the blood, forced by the action of the left ventricle into the arteries, was distributed to the body at large, and its several parts, in the same manner as it is sent through the lungs; impelled by the right ventricle into the pulmonary artery, and that it then passed through the veins and along the vena cava, and so round to the left ventricle in the manner already indicated. Which motion we may be allowed to call circular, in the same way as Aristotle says that the air and the rain emulate the circular motion of the superior bodies; for the moist earth, warmed by the sun evaporates; the vapours drawn upwards are condensed, and descending in the form of rain, moisten the earth again; and by this arrangement are generations of living things produced; and in like manner too are tempests and meteors engendered by the circular motion, and by the approach and recession of the sun.

And so, in all likelihood, does it come to pass in the body, through the motion of the blood; the various parts are nourished, cherished, quickened by the warmer, more perfect, vaporous, spiritous, and, as I may say, alimentive blood; which, on the contrary, in contact with these parts becomes cooled, coagulated, and, so to speak, effete; whence it returns to its sovereign the heart, as if to its source, or to the inmost home of the body, there to recover its state of excellence or perfection. Here it resumes its due fluidity and receives an infusion of natural heat – powerful, fervid, a kind of treasury of life, and is impregnated with spirits, and it might be said with balsam; and thence it is again dispersed; and all this depends on the motion and action of the heart.

The heart, consequently, is the beginning of life; the sun of the microcosm, even as the sun in his turn might well be designated the heart of the world; for it is the heart by whose virtue and pulse the blood is moved, perfected, made apt to nourish, and is preserved from corruption and coagulation; it is the household divinity which, discharging its function, nourishes, cherishes, quickens the whole body, and is indeed the foundation of life, the source of all action. But of these things we shall speak more opportunely when we come to speculate upon the final cause of this motion of the heart.

Hence, since the veins are the conduits and vessels that transport the blood, they are of two kinds, the cava and the aorta; and this not by reason of there being two sides of the body, as Aristotle has it, but because of the difference of office; nor yet, as is commonly said, in consequence of any diversity of structure, for in many animals, as I have said, the vein does not differ from the artery in the thickness of its tunics, but solely in virtue of their several destinies and uses. A vein and an artery, both styled vein by the ancients, and that not undeservedly, as Galen has remarked, because the one, the artery to wit, is the vessel which carries the blood from the heart to the body at large, the other or vein of the present day bringing it back from the general system to the heart; the former is the conduit from, the latter the channel to, the heart; the latter contains the cruder, effete blood, rendered unfit for nutrition; the former transmits the digested, perfect, peculiarly nutritive fluid.

SOURCE: Willius and Keys. *Classics of Cardiology* (New York, 1941), 36–9 and 47–9

22 Pierre Grassendi
'A NICE OBSERVATION OF THE PERVIOUSNESS
OF THE SEPTUM OF THE HEART'

I SHALL describe what I myself have seen.

While I was residing in Aix, whenever a dissection was being performed I was present frequently in the anatomical amphitheatre. Now for many years I had observed invariably that dissectors, taking the heart in their hands, would test the perviousness of its septum with a blunt instrument which they call a spatula, and would conclude, as physicians have concluded, that the transmission of blood from the right chamber to the left must occur by insensible transudation.

Now when this problem came to be discussed by the professors of anatomy, eight years ago, there came among the disputants a diligent surgeon, Payanus by name, who wanted to demonstrate to us onlookers that the facts were otherwise. So, taking up the spatula, he undertook to penetrate the mediastinum of the heart. But he did not attempt to push the instrument straight through, as the others had done, but having introduced its tip (for the tissue of the septum presents a thousand little openings) pushed onward with utmost gentleness, turning the instrument with the greatest patience up and down and from side to side, seeking always a farther ingress. And at last the instrument was seen entering the left chamber. But then, because we alleged that he had made an artificial opening, he himself requested one of us to incise the septum down to his instrument, with a sharp scalpel. When the incision had been made we found that no tissue anywhere had been injured, and we saw that only the meatus, or canal, notwithstanding the fact that it was a very winding passage, was lined with a very thin and glistening membrane.

These, then, are the passages which I mentioned to Fludd (who acknowledged clearly that he had been ignorant of their existence too); and I said that since they really do exist they cannot be without some function, and therefore it should be evident that there is a real percolation of blood from the right chamber into the left. And I maintained that the arterial blood was derived in this way.

Indeed, it seems probable that the more subtle part of the blood is, so to speak, sucked through this septum, or forced through by compression. But the grosser part of the blood, with the heavy vapours which it contains, enters the patent pulmonary artery and pervades and nourishes the tissues of the lungs. Then, after expiration has carried off the heavy breath and the heavy vapours, the more subtle residue of this blood is gathered into the pulmonary vein so that, together with the purer air which was inhaled in breathing, it may flow into the left ventricle, either drop by drop, as the general notion says it must, or in large spurts, as Harvey's opinion seems to have it.

SOURCE: as Document 21, 84–5

23 J. B. Van Helmont
'THE SECRETS OF PARACELSUS'

In Index or Table of the Secrets of *Paracelsus*; is, First of all, the Tincture of Lile, reduced into the Wine of Life, from an untimely mineral Electrum or general composure of Mettals; one part whereof is the first Metallus, but the other, the Essence of the Members.

And then follows *Mercurius Vitæ*, the off-spring of entire Stibium, which wholly sups up every Sinew of a Disease.

In the third place, is the Tincture of Lile, even that of Antimony, almost of the same efficacy or with that going before, although of less efficacy.

In the fourth place, is *Mercurius Diaphoreticus*, being sweeter than Honey, and being fixed at the Fire, hath all the Properties of the Horizon of Sol: for it perfects whatsoever a Physitian and Chyrurgion can wish for, in healing; yet it doth not so powerfully renew, as those Arcanums aforegoing.

His Liquor Alkahest is more eminent, being an immortal, unchangeable, and loosening or sol-giving water, and his circulated Salt, which reduceth every tangible Body into the Liquor of its concrete or composed Body.

The Element of Fire of Copper succeedeth, and the Element or Milk of Pearls. But the Essences of Gems and Herbs, are far less Arcanums than those aforesaid.

Lastly, the volatile Salts of Herbs, and Stones, do shew forth a precise particularity; neither do they reach unto the efficacy of Universal Medicines.

But His is Corollate, the which one alone, is purgative by Stool, cures the Ulcers of the Lungs, Bladder, Wind-pipe, Kidneys, by purging; so that it also utterly roots out the Gowt.

Indeed it is the Mercury of the Vulgar, from which, the Liquor Alkahest hath been once distilled, and it resides in the bottom, coagulated and powderable, being not anything increased, or diminished in its weight: From which Powder, the Water of the Whites of Eggs is to be cohobated until it hath attained the colour of Coral.

I praise the Lord of things, in an Abject or lowly Spirit; because he reveals his Secrets into the little Ones of this World, and doth always govern the Stern, least these his benefits should fall into the hands of the unworthy.

I have therefore discerned, that the Secrets of Paracelsus do take away Diseases; but that they reach not unto the Root of long Life.

I have also discerned, that Mineral Remedies, unto whatsoever the highest degree they are brought, yet that they are unfit for yielding Norishment unto the first constitutive Parts, because they reserve the middle Life of the concrete Bodies from whence they were extracted: For, for that cause, they never wholly lay aside a mineral Disposition; Yea, and therefore they depart from the tenour of long Life.

Yea, neither shall I ever be easily induced to believe, that the Phylosophers Stone can vitally be united with us, by reason of its exceeding immutable substance, which is incredibly fixed against the tortures of the Fire, being undissolvably homogeneal or simple in kind; that is, by reason of its every way impossi-

bility of separation, destruction, and digestion; so far is it from
conducing to long Life: Histories subscribe unto me, that none
who obtained that Stone, enjoyed a long Life; but that a short
Life hath befallen many, by reason of the dangers undergone
in labouring.

But, moreover, neither let Hucksters hope, that Meats which
do mightily nourish, will perform long Life: For although they
may afford strength unto those that are upon recovery; yet they
afterwards weaken them, being nourished: The which, Cæsar
also testifies: For the more tender Meats are easily consumed,
breed tender Flesh, and suffumigate or smoake the vital
Powers through their more greatly adult favour. But the
Studies of Physitians, are buisied about the delights of the
Kitchin, which they name the Dietary Part: for they have been
misled into errour, by thinking; that if Food of good Juice, and
tender, being administered in a due dose, doth profit those upon
recovery; they have thought also, that the more strong Persons,
being manifoldly nourished with the same Food, shall be raised
up into the highest increase of strength: For there is not a process made in feeding, as in Arithmetick, where ten Pounds lift
up nine; and by consequence, a hundred Pounds, ninety: But
he that eats very much, and drinks abundantly, shall not therefore become stronger than he that shall live more moderately:
For truly, Nature keeps not so much the proportions of Numbers, as the proportions of the Powers of things alterable
according to the Power of their own Blas.* However it is, at
least-wise, it succeeds with Physitians according to their desire:
Because plenty of venal Blood breeds Excrements, Physitians
are called for, and so they command the rules of Food at least-
wise to profit themselves, and they shorten the Life in those that
live medicinally, and miserably.

First of all, a Disease is a certain evil in respect of Life, and
although it arose from sin, yet it is not an evil like sin, from a
Cause of deficiency, whereunto a Species, Manner, and Order
is wanting: But a Disease is from an efficient seminal Cause,

* A word in Middle English meaning 'a breath', given an astrological meaning
here.

positive, actual, and real, with a Seed, Manner, Species, and Order. And although in the beholding of Life, it be evil; yet it hath from its simple Being, the nature of Good: For that which in its self is good, doth produce something by accident; at the position whereof, the faculties inbred in the parts, are occasionally hurt, and do perish by an indivisible conjunction.

Defects therefore there are, which from an external Cause, do make an assault beyond or besides the faculties of Life concealed in the parts; and they are from strange guests, received within, and endowed with a more powerful or able Archeus: And from hence they are the more exceeding in the importunity of times or seasons, quantities, and strength.

In the next place, there are occasional defects, which (seeing Good doth bring forth Evil by accident, and doth oft-times proceed from our own vital powers) are endowed with properties of their own, as it were their seminal Beginnings, therefore they immediately tend unto the vanquishing of our powers as their end: The which therefore, I elsewhere call, Diseases Potestative or belonging to our Powers. But neither is that a Potestative Being, which the Schooles do call A Disease by consent, and do think to be made by a collection or conjunction of Vapours: But a Potestative Being contains the government of a constrained faculty, as well in respect of the authority of Life, as of the diseasie Being it self; the which indeed is born by a proper motion, to stir up a Potestative Disease of its own order: Just as a Cantharides doth stir up a Strangury: And that also is done through a power of internal authority, and by the force of parts on parts. So an Apoplectical, or Epileptical Being, being as yet present in the Stomack, or Womb, shakes the Soul, yea and from thence transports the Brain, together with its attending powers, will they nill they, into its own service.

A Potestative Being therefore, doth not only denote a hurting of the Functions, but also a government of the part, and an occasioning force of a Diseasifying Being prorogued or continued on the subordinate faculties, as on the vassals of an Empire: It being all one also, whether the parts are at a far distance from each other, or whether they are near: For they are the due Tributes of Properties.

Yea truly, Hippocrates first insinuated, that Diseases are to be distinguished by their Inns and Savours: And I with his Successors had kept this tenor. But that Old Man being as it were swollen with fury, presaged of the future rashnesses of the suceeding Schools, and precisely admonished them, That they should not believe, that Heats, Colds, Moistures, Sharpnesses, or Bitternesses, were Diseases: But Bitter, Sharp, Salt, Brackish, etc. it self. But he sung these things before deaf or bored ears: For truly, the long since fore-past Ages, being inclined unto a sluggishness of enquiring, and an easie credulity, snatched up the scabbed Theorems of Heats and Colds, and subscribed unto them by reason of a plausible easiness, and bid Adieu to their Master; who having supposed that Diseases were to be divided according to their Innes, divided our body into three ranks, to wit, into the solid part containing or the vessel it self, into the thing contained, or liquid part; and into the Spirit, which he said was the maker of the assault. The which indeed is an Airy or Skiey, and Vital Gas, and doth stir up in us every Blas, for whether of the two ends you will. Which division of Diseases, although he hath not expressly dictated, yet he hath sufficiently insinuated the same: For he wrote onely a few things, and all things almost which are born about, are supposed to be his. And therefore I wish that posterity had directed the sharpnesses of their Wits, according to the mind of that Old Man; Per-adventure, through Gods permission, they had extracted the understanding of the Causes of Diseases: But they afterwards so subscribed unto the Authority of one Galen, that they, as it were slept themselves into a drousie Evil, being afrightned while they are awakened by me. But in the Title of Causes, I understand, in the very inward or pithy integrity of Diseases, the matter being instructed by its own proper efficient Cause, to be indeed the inward, immediate Cause, and to arise from a vital Beginning.

SOURCE: J. B. Van Helmont. *Oratrike, or Physick Refined* (1622),

804–6, 'The Arcanus or Secrets of Paracelsus', translated by John Chandler

24A Sir Kenelme Digby
'WEAPON SALVES'

Behold now, Sirs, the genealogie of the Powder of Sympathy in this part of the World, with a notable History of a cure performed by it. It is time now to come to the discussion, which is, to know how it is made. It must be avowed that it is a marveilous thing, that the hurt of a wounded person, should be cured by the application of a remedy put to a rag of cloth, or a weapon at a great distance. And it is not to be doubted, if after a long and profound speculation, of all the œconomy and concatenation of naturall causes, which may be adjudged capable to produce such effects, one may fall at last upon the true causes, which must have subtill resorts and means to act. Hitherto they have been wrapped up in darknesse, and adjudged so inaccessible, that they who have undertaken to speak or write of them, (at least those whom I saw) have been contented to speak of some ingenious gentilenesse, without diving into the bottom, endeavouring rather to shew the vivacity of their spirit, and the force of their eloquence, than to satisfy their Readers, and Auditors, how the thing is really to be done. They would have us take for ready mony, some terms which we understand not, nor know what they signifie. They would pay us with conveniences, with resemblances, with Sympathies, with Magnetical virtues, and such terms, without explicating what these terms mean. They think they have done enough, if they feebly perswade any body that the businesse may be performed by a natural way, without having any recourse to the intervention of demons, and spirits: but they pretend not in any sort to have found out the convincing reasons, to demonstrate how the thing is done.

Sirs, if I did not hope to gain otherwise upon your spirits; I say, that if I did not believe, that I should be able to perswade you otherwise than by words, I would not have undertaken this enterprize: I know too well such a design requires a great fire, & vivacity of conceptions, volubility of tongue, aptnesse of expressions to insinuate as it were by surprisal, that which one cannot carry away by a firme foot, and by cold reasons, though solid. A Discourse of this nature ought not to attend a stranger, who finds himself obliged to display his sense in a language, wherein he can hardly expresse his ordinary conceptions. Nevertheless, these considerations shall not deterre me from engaging my self in an enterprize, which may seem to some much more difficult than that which I am now to performe, *viz.* to make good convincing proofs, that this Sympathetical cure may be done naturally, and to shew before you eyes, and make you touch with your finger how it may be done.

SOURCE: *A late Discourse made in a Solemne Assembly of Nobles and learned men at Montpellier in France; by Sir Kenelme Digby, Knight, etc. Touching the cure of wounds by the powder of sympathy,* R. White's translation (London, 1658), 14–17

24B Sir Kenelme Digby
'CHOICE AND EXPERIMENTED RECEIPTS'

For Deafness

Take Oyl of Bitter-Almonds, Oyl of Nardish, *Ana* six drams, Juyce of Onions, Juyce of Rue, *Ana* two drams, Black Hellebory half a scruple, Coloquint half a dram, Oyl of *Exceter* two drams; boyl this till the juyces be consumed; then take Wool, dip it in, and put it in the Ear.

Another for the same

Take of wild Mint, mortifie and squeeze it in the hand till it rendreth juyce; then take it with its juyce, and put it in the Ear, change it often; this will help the Deafness, if the person hath heard before.

For *Deafness* through Cold and humors clogging and be-

numing the Ears, causing sometimes pain in them; is, to drop
into the affected Ear one drop (no more) of Oyl or Quintessence
of Rosemary (which will not burn or pain them) and after it is
soaked in (by lying with that Ear up) stop the Ear slightly with
Cotton or black Wooll dipped in the Spirit of Rosemary. Repeat
the drop after a day still as often as you find need.

For the Small-pox

Take two or three grains of Saffron, and dry it well by the fire,
then make a *Nodulus* of it in fine linnen, and infuse and press it
in Posset-ale, Mace-ale or white Wine, till all the tincture and
vertue be drawn out: give that to the Patient, and keep him
warm: If he have soreness in the Throat, do thus: Boyl a
quarter of a spooneful of dryed Saffron (in a *Nodulus*) in half a
pint of Milk till it be very yellow; in this, boyl a broad Stag of
Linnen till it be throughly tincted, and put it warm and moist
under the Throat, as if you pinned it to keep on a Coif: When
this Stag is cold and dry, take it off, and put on a new one, to
which end you must have at least two, that one may heat in the
Milk, whilst the other is pinned about the Throat: This will
certainly take away all the pain of the Throat in six or eight
hours: you must not use Oyntment or grease to anoint the
Scabs, but only plain ordinary *Unguentum Album*, when the
Pustules begin to dry; and this hath preserved all my Children
from any marks.

*To drive the venemous vapours from the Heart and head in the Small-
Pox, Measles, etc. with great success with a familiar Julep or Emulsion,
take the following.*

Take Seeds of Citron one ounce, seeds of *Card. Bened.* one ounce
and an half: beat them well, and draw out all their Pulp with
two pounds of some fit Cordial-water, as of Scabious *Card.
Bened.* Mary-gold, or the like; and sweeten it with two or three
ounces (*q.s.*) syrup of Citron: Drink of this as often as you have
a mind, to drink a reasonable glass-full; to make it more
Alexipharmacal to take now and then, make a Julep containing

Confect. Hyacinthi or of Alkerms and Treacle-water, and dia-phoretick *Antimony*, and prepared Pearl, and what else you think fit, and put a spoonful or too of this into a draught of the Emulsion.

To prevent Marking in the Small-Pox.

This is a certain remedy for all Scars of the Small-pox, and to take away all the sharpeness and pain of the humor that breaks out in the Face; Do thus;

As soon as ever the Small-pox are come out in the Face, even at their first forming and appearing evidently to be the Pox (by their manifest coming out.) Take Oyl of Sweet-Almonds newly drawn by expression without fire, and with a feather or other fit means anoint over all the face, going over it with the feather severall times, (five or six) that you may be sure every part of it is thoroughly well humected, and that it runneth down backwards by both Ears, as the party lyeth on their back to be anointed: Then take a book or two, (as many as needs) of beaten Gold (of the purest) and lay it on all over the face, leaving not a speck of it uncovered with Leaf-gold; and rather lay two leafs, one upon another, then to leave any part bear; for wheresoever there is a defect of Gold, there will be a scar; The Leafs will soon fasten on, by the drying of the Oyl, and so grow to a hard crust, which will shale off after ten or twelve days, when the scabs are dry and shale off, and there will be no marks at all of the Sores; and in the mean time the Patient will feel no pain upon his Face, nor is any thing to be done to his Face after the Gold is layed on: This hath been tryed upon many with perfect success. Remember, how soveraign Gold is in curing and taking away the Acrimony of Ulcers: The Patients Eyes must not be covered with Gold nor anoynted with Oyl, but you must put Oyl and Gold upon the lids as close as you can to the hairs; so that neither of them get into the Eye, which is to be preserved with other fit things, as Saffron, &c.

To prevent pitting in the Small-pox, boyl Cream to an Oyl, and with that anoynt the wheals with a feather as soon as they begin to dry, and keep the Scabs always moist herewith: Make your face be anointed almost every half hour.

An Experimented Remedy for the Falling-Sickness.

Take *Cran. Human.* Pairings of Nails of Man, *Ana* two ounces, reduce this to fine Powder, and grind it upon a Marble-stone; then take of *Polypod. Querci, Viscus Querci, Viscus Corilium, Viscus* of Tiliot, *Ana* two drams, Peony root dryed half an ounce; beat this all into fine Powder; then take six ounces of fine Sugar, boyl it to the consistence of Rose-Sugar; then mix all the Powders with it, and let them well incorporate over the fire, stirring them well together, then take it from the fire, and make it into little Tablets, of which give one in the morning fasting, and another two hours after dinner, and one two hours after supper. Continue this whiles the Tablets lasteth.

SOURCE: Sir Kenelme Digby. *Choice and Experimented Receipts* (1668), 54–61

25 Malpighi
'ABOUT THE LUNGS'

There were two things which, in my epistle about observation on the lungs, I left as doubtful and to be investigated with more exact study.

(1) The first was what may be the network described therein, where certain bladders and sinuses are bound together in a certain way in the lungs.

(2) The other was whether the vessels of the lungs are connected by mutual anastomosis, or gape into the common substance of the lungs and sinuses.

The solution of these problems may prepare the way for greater things and will place the operations of Nature more clearly before the eyes. For the unloosing of these knots I have destroyed almost the whole race of frogs, which does not happen in that savage Batrachomyomachia of Homer. For in the anatomy of frogs, which by favour of my very excellent colleague, D. Carolo Fracassato, I had set on foot in order to become more certain about the membranous substance of the lungs, it happened to me to see such things that not undeservedly I can better make use of that (saying) of Homer for the present matter –

'I see with my eyes a work trusty and great.'

For in this (frog anatomy), owing to the simplicity of the structure, and the almost complete transparency of the vessels which admits the eye into the interior, things are more clearly shown, so that they will bring the light to other more obscure matters.

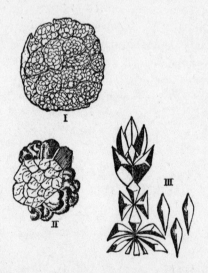

Fig. I. Outermost piece of dried lung showing the rete.

Fig. II. Interior vesicles and sinuses sketched with portion of the interstitium in the upper part. The beginning and complete prolongation could not be exhibited to the eye by the picture.

Fig. III. Adaptation over the trachea and the pulmonary vessels which also, parted from their usual site, are shown for easier understanding.

In the frog, therefore, the abdomen being laid open lengthwise, the lungs, adhering on each side to the heart, come forth.

They are not slack as in other animals, but remain tense for the animal's requirements. They are nothing more than a membranous bladder, which at first sight seems to be spattered with very small spots, arranged in order after the fashion of the skin of the dogfish – commonly called Sagrino. In form and surface protuberances it resembles the cone of a pine: but internally and externally a certain texture of vessels diversely prolonged is connected together, which, by the pulse, by contrary movement, and the insertion of the vein, are pulmonary arteries. In the concave and interior part of this (bladder) it almost fades into an empty space devoted to the reception of air, but it is not everywhere smooth but is interrupted by the occurrence of alveoli. These are produced by membranous walls raised to a little height. They are not all of this shape, but when the walls are produced out in length and width and connected together, the bays (sinuses) are formed almost into hexagons; and bent at the corners of the sinuses the membrane is extended a little as an infundibulum is constituted; and thus the lungs of the smaller frogs are fashioned. But in those which are larger, the walls are raised higher, and from the middle of the enclosed floor three come out very visibly increasing. The partitions in the smaller frogs are almost unobservable, but those in the bigger ones are bound into three other sinuses as they divide the greater sinus very much. The area, or the floor of the sinuses, admits the vessels spoken of above, and the artery itself sometimes ends inconspicuously, fork-like in the middle, but further on is spread out at the greater passage and sometimes manifestly produces another branch, but the vein glides down the inner slopes of the walls and is mingled with these, and, the branches having been sent down through the walls, at length runs into the area.

Observation by means of the microscope will reveal more wonderful things than those viewed in regard to mere structure and connection: for while the heart is still beating, the contrary (i.e., in opposite directions in the different vessels) movement of the blood is observed in the vessels, – though with difficulty, – so that the circulation of the blood is clearly exposed. This is more clearly recognized in the mesentery and in the other greater veins contained in the abdomen.

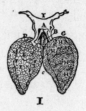

I

II

Fig. I. Showing lungs of frogs with trachea attached. (A) Larynx, which is semi-cartilaginous. (B) Rima, which is accurately closed and opened at the animal's need. Air being enclosed, it keeps the lungs expanded. (C) Site of the heart. (D) External part of the lung. (E) Prolonged rete of the cells. (F) Prolongation of the pulmonary artery. (G) Concave part of the lung divided through the middle. (H) Prolongation of the pulmonary vein running through the apices.

Fig. II. Containing the most simple cell without the intermediate walls (magnified). (A) Interior floor of the cell. (B) Parietes separated and bent. (C) Trunk of pulmonary artery with attached branches, as if ending in a network. (D) Trunk of pulmonary vein wandering with its branches over the slopes of the walls. (E) Vessel in the bottom and corners of the walls with the ramifications of the rete continued.

Thus, by this impulse, the blood is driven in very small (streams) through the arteries like a flood into the several cells, one or other branch clearly passing through or ending there. Thus the blood, much divided, puts off its red colour, and, carried round in a winding way, is poured out on all sides till at length it may reach the walls, the angles, and the absorbing branches of the veins. The power of the eye could not be extended further in the opened living animal, hence I had believed that this body of the blood breaks into the empty space, and is collected again by a gaping vessel and by the structure of the walls. The tortuous and diffused motion of the blood in divers directions, and its union at a determinate place offered a handle to this. But the dried lung of the frog made my belief dubious. This lung had, by chance, preserved the redness of the blood in (what afterwards proved to be) the smallest vessels, where by means of a more perfect lens, no more there met the eye the points forming the skin called Sagrino, but vessels mingled annularly. And, so great is the divarication of these vessels as they go out, here from a vein, there from an artery, that order is no longer preserved, but a network appears made up of the prolongations of both vessels. This network occupies not only the whole floor, but extends also to the walls, and is attached to the outgoing vessel, as I could see with greater difficulty but more abundantly in the oblong lung of a tortoise, which is similarly mebranous and transparent. Here it was clear to sense that the blood flows away through the tortuous vessels, that it is not poured into spaces but always works through tubules, and is dispersed by the multiplex winding of the vessels. Nor is it a new practice of Nature to join together the extremities of vessels, since the same holds in the intestines and other parts; nay, what seems more wonderful, she joins the upper and the lower ends of veins to one another by visible anastomosis, as the most learned Fallopius has very well observed . . .

From these things, therefore as to the first problems to be solved, from analogy and the simplicity which Nature uses in all her operations, it can be inferred that that network which formerly I believed to be nervous in nature, mingled in the bladders and sinuses, is (really) a vessel carrying the body of blood thither or carrying it away. Also that, although in the lungs of perfect animals the vessels seem sometimes to gape and end in the midst of the network of rings, nevertheless, it is likely that, as in the cells of frogs and tortoises, that vessel is prolonged further into very small vessels in the form of a network, and these escape the senses on account of their exquisite smallness.

Also from these things can be solved with the greatest probability the question of the mutual union and anastomosis of the vessels. For if Nature turns the blood about in vessels, and combines the ends of the vessels in a network, it is likely that in other cases an anastomosis joins them; this is clearly recognized in the bladder of frogs swollen with urine, in which the above described motion of the blood is observed through the transparent vessels joined together by anastomosis, and not that those vessels have received that connection and course which the veins or fibres mark out in the leaves of nearly all trees.

To what purpose all these things may be made, beyond those which I dealt with in the last letter concerning the pulmonary mixing of the blood, you yourself seemed to recognize readily, nor is the opinion to be lessened by your very famous device, because by your kindness you have entrusted me with elaborate letters in which you philosophised subtly by observing the strange portents of Nature in vegetables, when we wonder that apples hang from trunks not their own, and that by grafting of plants the processes have produced bastards in happy association with legitimates. We see that one and the same tree has assumed diverse fashions in its branches, – while here the hanging fruits please the taste by a grateful acidity, there they fulfill every desire by their nectar-like sweetness, and you furnish credibility to the truth at which you wondered when in Rome, that the vine and the jasmine had come forth from the bole of the Massilian apple. He who cultivated the gardens with a light inserted fork made these clever things with bigger branches, and he taught the unreluctant trees the bringing forth of divers things. About this matter Virgil in the Georgics fitly sang:

'They ingraft the sprout from the alien tree
And teach it to grow from the moist inner bark.'

You lay bare the secret of this wonderful result by your philosophising method, for we might consider the acid juice of the Massilian apple sweetens to the nature of pure wine as far as the particles of that juice may run through the small openings of the trunk proper, but not in the same way can they come up into the continued tubules of the vine. Here, stirred by their own motion, and torn away beyond their usual order by the impulse of those following after, and broken up, they must conform themselves to the superinduced form of the passage, and

126

put on the new nature by which the vine or jasmine is brought forth. Nature pursues a like mode of operation in the lungs, for the turbid blood returns from the ambit of the body, widowed elsewhere of particles, to which a new humour from the subclavian vein is added to be perfected by the further action of Nature. This happens in order that it may be arranged and prepared into the nature of particles of flesh, bone, nerve, etc., while it enters the myriad vessels of the lungs. It is conducted into divers very small threads. Thus a new form, situation, and motion is prepared for the particles of the blood, from which flesh, bone, and spirits may be formed. The trustworthiness of your saying is increased by the like structure of the seminal vessels as if a certain nutrition of the living animal were also its regeneration.

SOURCE: as Document 21, 92–7

FLORA AND FAUNA: TAXONOMY AND TOPOGRAPHY

By the end of the seventeenth century Europe was overwhelmed by the collecting and cataloguing mania that had characterised much of the preceding 100 years. Antiquarian collections had long been in vogue; they were now joined by increasing numbers of collections of flora and encouraged by the massive works of compilation pouring from the presses. In France the Académie Royale des Sciences produced volumes of detailed descriptions of flowers and animals; in England Ray, Lister, Plot, Lhwyd and others were assiduous editors, often keen to pioneer with Ray some new taxonomic principle which they could illustrate by reference to their profusion of knowledge and material. Lhywd as keeper of the Ashmolean Museum in Oxford, Plot as the great local antiquary and naturalist, and Ray as a systemiser and natural theologian, fed the increasingly important study of natural philosophy with their great works.

Reports flowed into Western Europe from the colonies. Details of new stones, minerals, plants, animals, and rarities were brought back in sketches and reports by gentlemen travelling from country to country keeping commonplace books. Travellers and local naturalists attempted to put into practice the rhetoric of looking at things themselves, away from the direction of books and the schoolroom. But books there were in profusion as well. Ray's Observations *show the kind of interest that the intelligent and learned gentlemen would pursue in a foreign city or countryside. Ray remarks both upon the urban landscape and upon the agricultural activities taking place in the neighbouring countryside. He is interested both in the buildings and in the plants that he discovers in the fields and mountains (26).*

Swammerdam's Book of Nature *provides us with but one example of the popular genre of botanical and biological studies. Swammerdam's work on insects (27) is of a high quality, representing a set of intelligent and precise observations from which he drew conclusions both natural and theological. It is interesting to note the way in which he introduces final cause apologia into his study of natural philosophy. He sees the way in which parts of animals and plants are well adapted to the functions they have to perform and to the lifestyle they have to evolve. This he sees as being conclusive evidence of the all-wise creator making them for that purpose. This argument was perhaps the most popular theological argument of the late seventeenth century and Swammerdam is entirely characteristic in his handling of it.*

26 John Ray
'TRAVELS'

Geneva is pleasantly seated at the lower end of the *Lacus Lemanus*, now called *Genffer-zee* or the lake of *Geneva*, upon a hill side respecting the lake; so that from the lake you have a fair prospect of the whole Town. It is divided by the river *Rhodanus* or *Rhosne* into two parts, which are joyned together by two wooden bridges, one of which hath on each side a row of houses after the manner of *London*-bridge, only they are low. The two principal and indeed only considerable streets in the Town are the low street [*Rüe bas*] which runs along by the river and lakes side: and the high street or great street [*Rüe grand*] which runs up the hill. The City is indifferently strong, and they have lately been at great expences to fortifie it *alla moderna* with ramparts and bastions of earth. Though it be but small yet is it very populous, being supposed to contain 30000 souls. St. *Peter's*

Church, formerly the Cathedral, is handsome and well-built, and in it is a stately monument for the Duke of *Rohan*. The Citizens are very busie and industrious, subsisting chiefly by trading: the whole territory of this Republic being not so great as some one Noblemans estate in *England* for extent of land. All provisions of victuals are very plentiful and cheap at *Geneva*, especially milk-meats, the neighbouring mountains feeding abundance of cattel.

The tops of *Jura*, *Saleve* and other high mountains of *Savoy*, *Daulphiny* and the *Alps*, where they are bare of wood, put forth very good grass so soon as the snow is melted off them, which usually is about or before mid-May. And then the Countrey people drive up their cattel to pasture, and feed them there for three months time. Upon these hill tops they have heer and there low sheds or dairy houses, which serve the men to live in and to make their butter and cheese in, so long as they keep their beasts above. The men I say, for they only ascend up thither and do all the dairy work, leaving their wives to keep house below; it being too toilsome for them to clamber up such high and steep hills. By reason of these cotes it is very convenient simpling upon the mountains, for if a man be hungry or thirsty he may soon find relief at one of them. We always found the people very kind and willing to give us such as they had, *viz*. brown bread, milk, whey, butter, curds, *etc.* for which we could scarce fasten any mony upon them.

For the temper of the air in respect of heat and cold *Geneva* I think is very like *England*, there being no great excess of either extreme. The City is well governed, vice discountenanced, and the people either really better or at least more restrained then in other places: Though they do take liberty to shoot and use other sports and exercises upon the Lords day, yet most of their Ministers disallow it and preach against it.

Our long stay heer at *Geneva*, and that in the proper season for simpling, gave us leisure to search for and advantage of finding many *species* of plants in the neighbouring fields and mountains, of which I shall heer present the Reader with a catalogue: Such as are native of *England* are in the *Roman* letter.

In Colle la Bastie *dicto &c Sylvis clivosis ad Rhodani ripas.*

Colutea scorpioides: Melissophyllon Fuchsio flore albo atque etiam vario: *Lilium flore nutante ferrugineo majus.* J. B. *Monophyllon* Ger. *Orobus Pannonicus* I Clus. *Hepaticum trifolium* Lob. Frangula; *Chamædrys falsa maxima,* &c. J.B. *Bellis sylvatica* J.B. *Trifolium purpureum majus folio & Spica longiore* J.B. Orobus *sylvaticus viciæ foliis* C.B. *Tithymalus cyparissias* J.B. *Tithymalus non acris flore rubro.* J.B. *Horminum luteum* sive *Colus Jovis*; Aquilegia vulgaris.

SOURCE: J. Ray. *Observations Topographical, Moral and Physiological* (1673), 434–6

27 J. Swammerdam
'ON INSECTS'

Of the Life of the Vermicle or worm of the ephemerus, when out of the egg; and of its food. It is very worthy of notice, that these Vermicles or Worms never, or but very rarely, are observed to swim at the bottom of the river, or even in the middle of the water. They can indeed swim very swiftly, and move and throw themselves easily into serpentine windings in the water, whilst their head is bent sometimes up and sometimes down; the rest of the body advancing with the like twisting convolution and serpentine motions. But, notwithstanding they have this in their power, they are always found near the banks of rivers, and they live there in the most quiet parts. The more mud there is in the bottom, out of which they first rise, the greater number of these Worms is usually found. But you can very rarely catch them lying on the mud or adhering to it, but they live within the mud or clay itself in hollows made oblong and smooth. These are bored, not obliquely or downwards, but always parallel to the horizon: therefore, Vander Kracht says right in Clutius, that these insects live in separate little cells.

As the Bees, therefore with wonderful and perhaps inimitable art form their habitations with wax; in like manner do the Worms of the Ephemerus make these hollow tubes . . . or long holes for their residence, and bore them in the mud, in proportion to the bulk of their bodies. Hence, when these Worms are expelled out of their holes, so that they must creep on the plain or smooth bottom, which does not support every part of their bodies, they immediately lose their ability to go forward, though they are even surrounded with water, and are able to sustain or bear themselves up by swimming. This I have experienced, when I had drawn a great number of these Worms out of their holes, in order to dissect them; they always fell on their backs, and, as if they were in a swoon, could not turn themselves again: whereas, on the contrary, when they are in their little holes or burrows, they can creep very quickly backwards and forwards, and move themselves every way as they have occasion. I observe that it is common to all kinds of Worms which live in these kind of cells or holes to be able to move very quickly into their retreats, and when they are taken out of them, to faint as it were away. This I have observed in the Worms which live in hollow trees, and also in those which are found in fruit, in the tubercles of the leaves, and in the galls or warts of plants. It is very worthy of observation, that the Cossus or Worm of the great Beetle, whenever it is taken out of its house, covers its whole body with a web, by the help of which it forms a new hole for itself in the wood; for it could by no means pierce or make a hole, unless it were provided with some kind of stay or support to lean against by pressing its body in that part, and finding a due resistance.

The bait or Worm of the Ephemerus is so weak when out of its hole or little tube, that if at any time it ceases to move, when swimming in the water, it immediately sinks to the bottom in confusion, and there lies on its back.

We are to remark further, that as soon as the Worms of the Ephemerus have issued out of their eggs, they prepare to build their cells or houses, which we have observed are long and horizontal hollow tubes or caverns made in the clay or mud. But they make these tubes by degrees larger and larger, according to the size of the body, so that by this means the full grown Worms are always found in larger . . . the young ones in smaller tubes.

The all-wise Creator has given them parts appropriated to this purpose; their two fore legs are formed in some measure, as they are in the Moles and Mole Crickets. These Worms have jaws likewise, which are provided with two teeth somewhat like the forceps or claws of crabs, and these serve very well to assist in making those holes in the mud.

Hence you will immediately see them piercing or boring, when they are thrown into a little mud mixed with water. If you do not give them a sufficient quantity of the mud, they will nevertheless continue to undermine what they have, at one time hiding their head, and at another their body, and afterwards their tail, attempting to prepare new cells.

The fishermen say, they are certain from experience that these Worms, when the water sinks from the brink or edge of the river, always bore holes through the mud into a lower and deeper part, and that they likewise go to higher places, when the water rises. This, I think, they are under a necessity of doing, since they have several air-pipes in their tracheæ, by the help of which they frequently draw new air, which is necessary to their life. This they could not do, if they were confined at too great a depth when the water rose higher.

I have often observed that when they were drawn out of their cells and put on the wet sand, they have chose rather to creep out of the water, than go to the bottom under the sand. This might possibly be owing to the want of mud, and the warmth of the water, which is probably injurious to them.

As to the food of these creatures, it is very difficultly discovered, unless by the dissection of them, which taught me that they live on clay or mud only. Whenever you open them, you will always find mud both in the stomach and in the small and great guts. These Worms are therefore in this respect like the Moth, which feeds on the same substance of which it makes its habitation.

SOURCE: J. Swammerdam. *The Book of Nature* . . . (1758 ed), 105–6

Part Three

JOURNALS AND SCIENTIFIC INSTITUTIONS

JOURNALS

From the mid-seventeenth century the combination of scholarly interchange and the interests of the leisured classes in matter of natural philosophy led to a marked quickening of the pace of scientific journalism and the institutionalising of views and interests. Journals flourished to bring to men of one country the views of their contemporaries, or the views of overseas correspondents unknown to the editor; they also increasingly existed to provide readily accessible information about scientific and literary affairs to men little versed in them, or with insufficient time or knowledge to study them in depth. It was an age of popularisation, an age of gentlemanly interchange across frontiers and narrow seas.

Each of the major European centres of scientific advance had its organ of publication, sometimes connected to the major scientific institution in a personal or formal manner. The Philosophical Transactions *was the organ of the English Royal Society, and the Académie des Sciences produced several official publications, and was able to vent some of the work that it wished to be published widely through the* Journal des Scavans. *There were also the Italian* Giornale dei Republica dei Leterati, *Pierre Bayle's* Nouvelles de la République des Lettres, *and the German* Acta Eruditorum.

The quality of the work submitted for publication was very varied, and the subject matter and method of treatment were extremely diffuse. A publication like the Philosophical Transactions *would devote as much space to the occurrence of a monstrous birth in a provincial town as it would to an argument over the mathematical complications of gravita-*

tional theory. *The two extracts printed below (combined in 28) concerning a sand flood in the county of Suffolk and defective sight and its correction are but two examples of the subjects which would excite the imagination and the interest of people writing for and reading the* Transactions.

The extract from the History of the Works of the Learned *(29) illustrates the growth in review journalism, providing comment upon Du Hamel's* History of the French Academy of Sciences. *The review is interesting both for the summary it gives of Du Hamel's account of the origins and growth of the Academy and for the form in which it is written, bringing learning to those with insufficient time to read long books in learned languages – in fact, to the average gentleman in the coffee house.*

28 Royal Society
'PHILOSOPHICAL TRANSACTIONS, 1668'

A curious and exact Relation of a Sand-floud, *which hath lately overwhelmed a great tract of Land in the County of* Suffolk; *together with an account of the Check in part given to it; Communicated in an obliging Letter to the* Publisher, *by that Worthy Gentleman* Thomas Wright Esquire, *living upon the place, and a sufferer by that Deluge.*

Sir, I beg your pardon, that I have not made an earlier return to the Letter, by giving you the account, you required of those prodigious *Sands,* which I have the unhappiness to be almost buried in, and by which a considerable part of my small fortune is quite swallow'd up. But I assure you, my silence was not the result of any neglect, but rather of my respects to you, whose employments I know are too great to suffer you often, *vacare nugis.* The truth is, I suspended the giving you any trouble, till I was put into some capacity of answering the whole Letter, as well concerning those few Improvements, this part of the Nation has made in *Agriculture,* as these wonderful *Sands,* which although they inhabit with and upon me, and have not yet exceeded one *Century,* since they first broke prison, I could not without some difficulty trace to their Original. But I now find it to be in a Warren in *Lakenheath* (a Town belonging to the *Dean* and *Chapter* of *Ely*, distant not above 5 miles, and lying *South-west* and by *West* of this place) where some great Sand-hills (whereof there is still a remainder) having the *Superficies*, or sword of the ground (as we call it) broken by the impetuous *South-west* winds, blew upon some of the adjacent grounds: which being much of the same nature, and having nothing but a thin crust of barren earth to secure its good behaviour, was soon rotted and dissolved by the other Sand, and thereby easily fitted to increase the Mass, and to bear it company in this strange progress.

At the first Eruption thereof (which does not much exceed the memories of some persons still living) I suppose, the whole Magazin of Sand could not cover above 8, or 10 acres of ground, which increas'd into a 1000 acres, before the Sand had travailed 4 miles from its first aboad. Indeed it met with this advantage, that till it came into this Town, all the ground, it past over, was almost of as mutinous a nature as it self, and wanted nothing but such a Companion to set it free, and to sollicit it to this new Invasion. All the opposition it met with in its Journy hither, was from one Farm-house, which stood within a mile and a half from its first source. This the Owner at first endeavour'd to have secur'd by force and building of Bulworks against the Assaults thereof; but this wing'd Enemy was not to be so oppos'd: which, after some dispute, the Owner perceiving, did not only slight the former Works, but all his Fences, and what else might obstruct the passage of this unwelcome guest, and in four years effected that by a compliance and Submission, which could never have been done upon other terms: In which he was so successful, as that there is scarce any footsteps left of this mischievous Enemy.

'Tis between 30 and 40 years, since it first reacht the bounds of this Town; where it continued for 10, or 12 years in the Out-skirts, without doing any considerable mischief to the same. The reason of which I guess to be, that its Current was then *down-hill*, which shelter'd it from those winds, that gave it motion. But that Valley being once past, it went above a mile (*up-*

hill) in two months time, and over-ran 200 acres of very good Corn the same year. 'Tis now got into the body of this little Town, where it hath buried and destroy'd divers Tenements and other Houses, and has inforc'd us to preserve the remainder at a greater charge than they are worth. Which doubtless had also perisht, had not my affection to this poor dwelling oblig'd me to preserve it at a greater expence than it was built: Where at last I have given it some Check, though for 4. or 5. years our Attaques on both sides were with so various success, as the Victory remain'd very ambiguous. For, it had so possest all our Avenues, as there was no passage to us but over two Walls of 8. or 9. foot high (which incompass'd a small Grove before my house, now almost buried in the Sand;) nay, it was once so near a conquest, as at one end of my house it was possest of my Yard, and had blown up to the Eves of most of my out-houses. At the other end it had broke down my Garden-wall, and stopt all passage that way.

But during these hard and various skirmeshes I observed, that that Wing of Sand, that gave me the assault, began to contract into a much less compass. For by stopping of it 4. or 5. years (what I could) with Furre-hedges, set upon one another, as fast as the Sand levell'd them (which I find to be the best Expedient to hinder its passage, and by which I have raised Sand-banks near 20 yards high) I brought it into the Circuit of about 8. or 10 acres: And then in one year by laying some hundreds of Loads of Muck and good earth upon it, I have again reduced it to *Terra firma*, have clear'd all my Walls, and by the assistance and kindness of my neighbours (who help'd me away with above 1500 loads in one month) cut a passage to my house through the main body thereof.

But the other end of the Town met with a much worse fate, where divers dwellings are buried or overthrown, and our Pastures and Meadows (which were very considerable to so small a Town, both for quantity and quality) over-run and destroy'd: And the branch of the River *Ouse*, upon which we border, (being better known by the name of *Thetford*- or *Brandon*-River, between which two Towns we lye,) for 3 miles together so fill'd with Sand, that now a Vessel with two load weight passeth with as much difficulty as before with 10. But had not the stream interpos'd, to stop its passage into *Norfolk*, doubtless a good part of that Country had ere now been left a desolate Trophy of this Conquering Enemy. For according to the proportion of its increase in these 5 miles, which was from 10. acre to 15 0. or 20 0; in 10 miles more of the same soil it would have been swell'd to a great vastness.

And now, Sir, I have given you the History of our Sands, I shall out of my respects to your design, (which I truly venerate, and should be glad to be subservient to in the meanest capacity) make this poor Essay towards a Discovery of a *Reason* and *Cause* of this strange Accident. Where the first thing observable to me is the quality and situation of the Country, in which this troublesome Guest first took his rise; which lyes *East-Nord-East* of a part of the great Level of the *Fenns*, and is thereby fully exposed to the rage of those Impetuous blasts, we yearly receive out of the opposite quarter: which, I suppose, acquire more than ordinary vigor by the winds passing through so long a Tract, without any check (which, when it has gone so far in triumph, makes its first assaults with the greater fury.) The other thing, that contributes to it, is, the extream Sandiness of the Soyl, the levity of which, I believe, gave occasion to that Land-story of the *Actions* that use to be brought in *Norfolk* for Grounds blown out of the Owners possession. But this County of *Suffolk* is more friendly in that particular, I having hitherto possessed great quantities of this *Wandring* land, without any scruple; which I should yet be glad to be ridd off without anything for the keeping, if the Owners would but do me kindness to fetch it away.

As to our *Georgicks*, they are so little the care and study of any Ingenious persons in these parts, that I am asham'd, I must be so breef upon a Subject so much every bodie's concern. The greatest matters that have been done, hath been by *Marling*: For, 50 load of *Marle* to an acre of dry barren lingy Heath make (as they say) a very great improvement both for *Corn*, *Turnips*, *Clover-grass*, *Nonsuch* and *Cole-seed*. Of the 3 first, I suppose, I

137

need to say nothing. But of the 2. last, (which are late Experiments) I have received a very good account from some *Norfolk* Gentle-men, one of whom the last year had of 7 acre of Nonsuch or Hopp-Clover 70 loads of Seed, besides a great crop of good Hay; which was twice as much worth as the best crop of Wheat in this Country. 'Tis sown (as the Common Clover) with Corn, and when it once takes, it will hold 4 times as long in the ground. About a bushel and an half soweth an Acre; and the Seed is now brought to 12s. the *Comb* (or 4 bushels) which was lately at 40s. The same Gentleman had the last year 10 *Combs* per acre of Coleseed upon a very dry heath (only improv'd by *Marling*) and was this year in expectation of a much greater crop, when I last saw him, I am, Sir, Your, &c.
Downham Arenarum, in *Suffolk*, July 6, 1668.

An Extract of a Letter concerning an Optical Experiment, conducive to a decayed Sight, communicated by a Worthy person, who found the benefit of it himself.

I am to acquaint you of an Experiment, if it may deserve that name, and not rather that of a Trifle; the matter of which is known to many, but un-applied (for ought I know) to such use as it affords. And the use is to my self of greater value, than you'l easily imagine, and I think, it may be equally profitable to many. Thus it is: you know, I have mourn'd for the loss of my eyes. I confess my unmanliness, that I have shed many tears in my study for want of them; but that was quite out of the way of recovering them. I know not, whether by standing much before a blazing Fire, or by writing often right before a bright Window, or what else might be the cause of this decay of my sight, who am not above 60 years of age. But I seem'd always to have a kind of thick smoak or mist about me, and some little black balls to dance in the air about my eyes, and to be in the case, as If I came into a room suddenly from a long walk in a great Snow. But so it was, I could not distinguish the Faces of my acquaintance, nor Men from Women in rooms that wanted no light. I could not read the great and black English Print in the Chuch-Bibles, nor keep the plain and trodden paths in Fields or Pastures, except I was led or guided. I received no benefit by any *Glasses*, but was in the case of those, whose decay by Age is greater than can be helped by *Spectacles*. The *fairest* Prints seemed through Spectacles like *blind* Prints, litle black remaining.

Being in this sad plight, what trifle can you think hath brought me help, which I value more than a great Sum of Gold. Truly, no other than this. I took Spectacles that had the largest Circles; close to the semi-circles, on the over-part, on both sides, I cut the bone; then, taking out the Glasses, I put black *Spanish* leather taper-wise into the emptied circles, which widen'd enough (together with the increasing wideness of the Leather,) took in my whole eye at the wider end; and presently I saw the benefit through the lesser taper-end, by reading the smallest Prints that are, as if they had been a large and fair Character. I caus'd a Glover to sow them with a double-drawn stitch, that they might have an agreeable roundness, and exclude all rayes of light. So I colour'd the Leather with Ink, to take off the glittering. And this was all the trouble I had, besides the practice and patience in using them. Only, finding that the smaller the remote orifice was, the fairer and clearer the smallest Prints appear'd; and the wider that orifice was, the larger Object it took in, and so required the less motion of my hand and head in reading; I did therefore cut one of these Tapers a little wider and shorter than the other; and this wider I use for ordinary Prints, and the longer and smaller for smallest Prints: These without any trouble, as oft as I see need, or find ease in the change, I alter. I can only put the very end of my little finger into the orifice of the lesser, but the same finger somewhat deeper, yet not quite up to the first joynt, I can insert into the orifice of the wider. Sometimes I use one eye, sometimes another, for ease by the change; for you must expect that the visual rayes of both eyes will not meet for mutual assistance in reading, when they are thus far divided by Tubes of that length.

The lighter the stuff is, the less it will cumber. Remember alwayes to black the inside with some black that hath no lustre or glittering. And you should have the Tubes so moveable, that you may draw them longer or shorter, allowing also (as was newly intimated) the orifice wider or narrower, as is found more helpful to them, that have need of them. To me it was not necessary, but I conceive it convenient, that Velvet or some gentle Leather should be fastned to the Tabulous part next the eyes, to shadow them from all the encompassing light.

I have already told you, that I found no benefit at all by any kind of Spectacle glasses, but I have not tryed, what Glasses will doe, if setled in these Tubes; having no need of them, I rest as I am. Now I should be heartily glad, if any of my friends should receive any aid or ease by such an obvious device (containing nothing but emptiness and darkness) as this is. And probably they may be more proper for some that are *Squint-eyed*, whose eyes doe interfere, and so make the object, as if you would write one line upon another, where, though both should be ever so fairly written, yet neither will be easily legible. Here *Squint-eyes* will be kept in peace, and at fair Law. Certainly it will ease them, that cannot well bear the light; and perchance it will preserve the sight for longer durance. If *N. N.* should find the benefit, as I do, he may thank you for the information, &c.

SOURCE: Royal Society. *Philosophical Transactions* (1669), 722–5 and 727–9

29 'The Works of the Learned'
REVIEW OF DU HAMELS 'HISTORY OF THE FRENCH ACADEMY OF SCIENCES'

Regiæ Scientiarum Academiæ Historia, &c. i.e. The History of the French Royal Academie of Sciences; in which, besides the Account of its Origin and Progress, and the various Dissertations and Observations for Thirty Years past, there are a great number of other Experiments and Inventions, both in Physicks and

Mathematicks, digested in Order. By *John Baptist Du Hamel*. Fellow and Secretary of the said Academie. At *Paris* 1698. 40, containing 411 Pages

The Learned of all Nations do certainly owe very much to the Author; but Posterity will be more indebted to him for giving us so clear and succinct an History of the Rise, Progress, Annual Acts and Debates, of so Illustrious an Academy, in this small Volume. He undertook this Province above 20 Years ago, by the Advice of the Famous Abbot *Bignon*; but being diverted by other Labours, and chiefly in Writing his *System of Antient and Modern Philosophy*, he was obliged to break off his Enterprize; and afterwards the Cruel War, which lately shook all *Europe*, the Author's growing Age, weak State of Health, and many other Impediments, retarded the Edition of this Work. However, the Publick has this Advantage by it, that whereas the Author design'd to have brought the History down only to 1692. he hath now brought it to 1697. He is of opinion, That this Work will be so much the more grateful to all the Lovers of Polite Learning, because many of those Books which have been publish'd both in the Name of the whole Academy, and in the Names of particular Members, ever since its first Foundation are scarce to be had; and here they have an Account of all those Books in their proper Time and Place. To this add, the incredible Variety of things, which cannot but charm the Reader, when the Author going through Philosophy in all its parts, comprehends in this Volume, so many famous and profitable Inventions for the use of Mankind. For he treats of Natural Philosophy, Chymistry, Botany, Anatomy, Geometry, Algebra, Mechanicks, Hydrostaticks, Dioptricks and Astronomy, which he principally takes notice of in this Book, because for the sake of this most Curious Science so useful to the Church and Commonwealth, the Academy it self was chiefly founded; and the Observatory built by the King's Liberality. The Author excuses himself, that he hath writ this History in Latin and not in French, because the Latin Tongue is so much despised by most People now, and that he was desir'd not only to consult the Advantages of the Learned in *France*, but also of those of other Nations, that did not understand French, seeing that which *Tully* said of old of the Greek Tongue, may be properly said now of the Latin in respect of the French. The Latin is read almost in all Nations, but the French only in its own Territories, and those but small.

The Work is divided into Four Books: the first gives the History of the Foundation of the Academy, and its Acts from 1666. to 1675. the second to 1684. the third to 1692. and the fourth to 1696.

As to the Foundation of it, he tells us, That Peace being concluded betwixt *France* and *Spain* in 1659, the most Christian King having turn'd all his Care and Thoughts to the Administration of his Government and to procure the Welfare of his People He thought fit to add the Splendor of Learning and Sciences, to the Glory of his Empire enlarged by so many Victories: To which end he thought it the best and safest Way, that Men eminent for Learning, should form themselves into a Society by the Consent of Publick Authority, and to confer and debate together upon their Inventions and thoughts, which he perceived would be much more profitable than if they laboured singly in the promoting of any one Science. And therefore he ordered the most Illustrious M. *Colbert*, whom he had chosen to be one of his chief Ministers, to bring this design to effect, which he himself had projected. M. *Colbert* having taken Advice with Learned and Prudent Men, resolved, That the Society should be formed of Men very well versed in Physicks and Mathematicks, but so as they should excel in one, more than in the other, without neglecting the rest however, for that excellent Person was of Opinion, That those Sciences were so strictly united, That 'twas scarcely possible, for any Man to excel in one that was not likewise well versed in the other.

Others persuaded him, that besides Mathematicians and Natural Philosophers, he should adopt into the Society other Learned Men, who had applied themselves to Polite Learning, and especially those that were well instructed in History; which Advice being approved, he appointed that the Mathematicians and Natural Philosophers, should meet separately on Wednesdays, and together on Saturdays, in that part of the Royal Library, which contained the Books of those Sciences. But those who applied themselves to History, he ordered to meet on Mondays and Thursdays, in that part of the King's Library which contained Historical Books, and that those who studied Polite Learning, should meet together on Tuesdays and Fridays. Then he ordered, that on the first Thursday of every Month, all those Societies should meet together; at which General Meeting (an Account being given by the Secretaries of the Academy) it should be lawful for every one to desire a Solution *ex tempore*; of those things that seem'd difficult to him but with this Caution, That if the Difficulties were greater than could be solved off-hand, then their Objections and Answers should be given in Writing, that the time of Meeting might not be spent in Unprofitable Contention.

But this first Constitution of the Academy was of no long duration, for in the very Commencement the Society of those who met for Illustrating History, was dissolved for certain Reasons: For since History, and especially that call'd Church-History, hath a strict Coherence with Questions in Divinity, and chiefly with those relating to the Government of the Church, and seeing from Matters of Fact oftentimes, Matters of Right are deduced, they were afraid that this Society of Learned Men, might offend those whom it was not safe to provoke. The Historical Society, being thus dissolved, he gives an Account how the Society of Polite Learning, was also separated from the Academy. Most of the Members of the said Society, being also Fellows of the French Academy, which they perceived to have now lost much of its first Splendor, to be almost desolate, and next door to ruine. They intreated M. *Colbert*, that he would shew the same Care, and Good-will towards that Ancient Academy, which he had been pleased to discover to the new one, and represented to him that there was no need of different Societies for the same things, especially seeing the same Persons did, in a manner compose both Academies. This Advice being no way displeasing to M. *Colbert*, he granted their

Request; and did so apply himself to restore and maintain this Academy, that he condescended to be one of the Fellows of the Society; and to honour them sometimes with his Presence; and by that means this School of Humanity being as it were torn from its own Body and confounded with the French Academy; The Academy of Natural Philosophy and Mathematicks only stood, and retaining its Original Strength, never lost any of its first Splendor.

Thus in June 1666. about 6 or 7 Mathematicians only began to meet, whose Names are as follow, M. *Carcavi, Hugens, de Roberval, Frenicle, Auzoult, Picard* and *Buat.* But it being proposed at first, that this Academy should also apply themselves to the Illustrating of Natural Philosophy. M. *Colbert* took care to chuse Men very well versed in the other Parts of Philosophy, who had read and seen much, and being addicted to no Sect of Philosophers, took delight in Sciences of all sorts. Therefore besides the Mathematicians above named, there were added to the Academy, M. *de la Chambre,* Physician in Ordinary to the King, M. *Perrault,* a Person who excell'd in Learning of all sorts, M. *Du Clos* and *Bourdelin,* extraordinarily versed in Chymistry, M. *Pecquet* and *Gayen,* skilful Anatomists, and M. *Marchant,* a Learned Botanist; and our Author was some Months before appointed to be their Secretary, to Write and Record what they proposed.

On the 22d of *Decemb.* 1666. those two Societies united and met together in the Hall of the King's Library, where they debated, Whether it was best that the Natural Philosophers, and Mathematicians should meet together, and so only form one Society; or, Whether they should meet a-part; and, because of the most strict Alliance there is betwixt the Physicks and Mathematics, they unanimously agreed, That they should not be separated, being thereunto incouraged by the Examples of great Men, who being very well skill'd in the Mathematicks, contributed much more to the Knowledge of Natural Philosophy, than those other Philosophers, that knew nothing of the Mathematicks. Such were *Galileus, Gassendus, Cabæus, Cartesius Honoratus Fabri,* and many others not needful to be named here.

Therefore they agreed, That both the Mathematicians and Natural Philosophers should meet twice every Week, upon the Mathematicks on Wednesdays, and Natural Philosophy on Saturdays, and that the Acts of the Academy should not be published without their own Order.

'Tis needless for us to specifie any of their Acts, our Author having exhibited all of them in a compendious manner. We shall only add what he hath said concerning the Observatory, or House for viewing the Stars, which the Most Christian King bountifully erected at his own Charge, for that Prince well knew. That Astronomy had not arrived to the height to which it was then arrived, but by frequent and accurate Observations; and that no new Additions could be made to it, but in the same manner.

The King chose a very convenient Place for it in the Suburb of St. *James*; for seeing that part is higher than the rest, it is less clouded with Smoak and Vapours, and therefore hath a free Aspect to all the Regions of the Heavens, especially towards the Antartick Pole, where the Observations of the Planets, are more frequently made. Our Author describes this Observatory thus:

> *This House is 80 Foot high, and the Foundations are dug as deep, because almost the whole Suburb and the Field next to it, is made hollow underneath; for they daily dig Stones from those Caves, either to build or repair the Houses; from whence it comes to pass, that the Descent of the Observatory into a Subteranean Cave, equals its height. There is a large Hole or Well, that reaches from the Bottom of this Observatory to the Top. 'Tis built of such hewen Stones, that there was no need of Mortar to fasten them.*

This Observatory begun to be inhabited by D. *Cassini*, and to be furnished with all sorts of Instruments in *September* 1671. Nor doth our Author say any thing more of its Form, because Monsieur *Perrault* hath accurately Described and Illustrated it with Four Figures, in his French Translation of *Vitruvius*, *l*. I. *c*. 2.

On the 21st of *August* 1690. the late King *James* visited this Observatory: Our Author largly recites what Observations were made then, and what Speeches past betwixt the said King and the Fellows of the Academy in *L.* III. *c.* 2. He adds, That the said King spoke with such Learning and Skill, that all that heard him admired and rever'd him. Some Years before, namely in 1682. on the 21st of *May, Lewis* XIV. himself, visited this observatory, and tooke notice of the Astronomical Instruments, and was informed by D. D. *Cassini, Picard* and *de la Heire*, of their Use, and what Observations were made by the help of them, like as in the Year before, on the 5th of *Dec.* 1681. he visited the Academy it self.

SOURCE: *The History of the Works of the Learned* (London, 1699), 216–20

THE IDEA OF THE ACADEMY

The scientific club or house of learning has many precedents. Probably the most influential, certainly in England, is to be found in Bacon's description of Salomon's House in 'New Atlantis', an institution to be devoted to research and study. Bacon's vision was part of a wider dream concerning the nature and purposes of political society, but the academy for the advancement of learning played an important rôle in his ideal commonwealth.

The extract below (30) hints at the general purpose and institution of Salomon's House, and details a few of the departments of the academy for the furthering of study, as well as detailing the categories of Fellows undertaking the work.

Bacon's vision can be contrasted with Swift's Grand Academy of Lagado (31). Gulliver is travelling through Balnibarbi and notes some details concerning the academy founded for furthering human knowledge, in the best satiric vein that Swift could muster. Swift lampoons all that was absurd, pointless and ridiculous in the trend of natural philosophy and academies for the improvement of industry and agriculture which he observed about him.

30 Francis Bacon
'NEW ATLANTIS'

Ye shall understand (my dear friends) that amongst the excellent acts of that king, one above all hath the pre-eminence. It was the erection and institution of an Order or Society which we call *Salomon's House*; the noblest foundation (as we think) that ever was upon the earth; and the lanthorn of this kingdom. It is dedicated to the study of the Works and Creatures of God. Some think it beareth the founder's name a little corrupted, as if it should be Solamona's House. But the records write it as it is spoken. So as I take it to be denominate of the King of the Hebrews, which is famous with you, and no stranger to us. For we have some parts of his works which with you are lost; namely, that Natural History which he wrote, of all plants, from the *cedar of Libanus* to the *moss that groweth out of the wall*, and of all *things that have life and motion*. This maketh me think that our king, finding himself to symbolize in many things with that king of the Hebrews (which lived many years before him), honoured him with the title of this foundation. I am the rather induced to be of this opinion, for that I find in ancient records this Order or Society is sometimes called Salomon's House and sometimes the College of the Six Days Works; whereby I am satisfied that our excellent king had learned from the Hebrews that God had created the world and all that therein is within six days; and therefore he instituting that House for the finding out of the true nature of all things, (whereby God might have the more glory in the workmanship of them, and men the more fruit in the use of them,) did give it also that second name . . .

The End of our Foundation is the knowledge of Causes, and secret motions of things; and the enlarging of the bounds of Human Empire, to the effecting of all things possible.

. The Preparations and Instruments are these. We have large and deep caves of several depths: the deepest are sunk six hundred fathom; and some of them are digged and made under great hills and mountains: so that if you reckon together the depth of the hill and the depth of the cave, they are (some of them) above three miles deep. For we find that the depth of a hill, and the depth of a cave from the flat, is the same thing; both remote alike from the sun and heaven's beams, and from the open air. These caves we call the Lower Region. And we use them for all coagulations, indurations, refrigerations, and conservations of bodies. We use them likewise for the imitation of natural mines; and the producing also of new artificial metals, by composition and materials which we use, and lay there for many years . . .

We have also sound-houses, where we practise and demonstrate all sounds, and their generation. We have harmonies which you have not, of quarter-sounds, and lesser slides of sounds. Divers instruments of music likewise to you unknown, some sweeter than any you have; together with bells and rings that are dainty and sweet. We represent small sounds as great and deep; likewise great sounds extenuate and sharp; we make divers tremblings and warblings of sounds, which in their original are entire. We represent and imitate all articulate sounds and letters, and the voices and notes of beasts and birds. We have certain helps which set to the ear do further the hearing greatly. We have also divers strange and artificial echos, reflecting the voice many times, and as it were tossing it: and some that give back the voice louder than it came; some shriller, and some deeper; yea, some rendering the voice differing in the letters or articulate sound from that they receive. We have also means to convey sounds in trunks and pipes, in strange lines and distances.

We have also perfume-houses; wherewith we join also practices of taste. We multiply smells, which may seem strange. We imitate smells, making all smells to breathe out of other mixtures than those that give them. We make divers imitations of taste likewise, so that they will deceive any man's taste. And in this house we contain also a confiture-house; where we make all sweet-meats, dry and moist, and divers pleasant wines, milks, broths, and sallets, far in greater variety than you have.

We have also engine-houses, where are prepared engines and instruments for all sorts of motions. There we imitate and practise to make swifter motions than any you have, either out of your muskets or any engine that you have; and to make them and multiply them more easily, and with small force, by wheels and other means: and to make them stronger, and more violent than yours are; exceeding your greatest cannons and basilisks. We represent also ordnance and instruments of war, and engines of all kinds: and likewise new mixtures and compositions of gun-powder, wildfires burning in water, and unquenchable. Also fire-works of all variety both for pleasure and use. We imitate also flights of birds; we have some degrees of flying in the air; we have ships and boats for going under water, and brooking of seas; also swimming-girdles and supporters. We have divers curious clocks, and other like motions of return, and some perpetual motions. We imitate also motions of living creatures, by images of men, beasts, birds, fishes, and serpents. We have also a great number of other various motions, strange for equality, fineness and subtilty.

We have also a mathematical house, where are represented all instruments, as well of geometry as astronomy exquisitely made.

We have also houses of deceits of the senses; where we represent all manner of feats of juggling, false apparitions, impostures and illusions; and their fallacies. And surely you will easily believe that we that have so many things truly natural which induce admiration, could in a world of particulars deceive the senses, if we would disguise those things and labour to make them seem more miraculous. But we do hate all impostures and lies: insomuch as we have severely forbidden it to all our fellows, under pain of ignominy and fines, that they do not shew any natural work or thing, adorned or swelling; but only pure as it is, and without all affectation of strangeness.

These are (my son) the riches of Salomon's House.

For the several employments and offices of our fellows; we have twelve that sail into foreign countries, under the names of other nations, (for our own we conceal;) who bring us the books, and abstracts, and patterns of experiments of all other parts. These we call Merchants of Light.

We have three that collect the experiments which are in all books. These we call Depredators.

We have three that collect the experiments of all mechanical arts; and also of liberal sciences; and also of practices which are not brought into arts. These we call Mystery-men.

We have three that try new experiments, such as themselves think good. These we call Pioners or Miners.

We have three that draw the experiments of the former four into titles and tables, to give the better light for the drawing of observations and axioms out of them. These we call Compilers.

We have three that bend themselves, looking into the experiments of their fellows, and cast about how to draw out of them things of use and practice for man's life, and knowledge as well for works as for plain demonstration of causes, means of natural divinations, and the easy and clear discovery of the virtues and parts of bodies. These we call Dowry-men or Benefactors.

Then after divers meetings and consults of our whole number, to consider of the former labours and collections, we have three that take care, out of them, to direct new experiments, of a higher light, more penetrating into nature than the former. These we call Lamps.

We have three others that do execute the experiments so directed, and report them. These we call Inoculators.

Lastly, we have three that raise the former discoveries by experiments into greater observations, axioms, and aphorisms. These we call Interpreters of Nature.

We have also, as you must think, novices and apprentices, that the succession of the former employed men do not fail; besides a great number of servants and attendants, men and women. And this we do also: we have consultations, which of the inventions and experiences which we have discovered shall be published, and which not: and take all an oath of secrecy, for the concealing of those which we think fit to keep secret: though some of those we do reveal sometimes to the state, and some not.

SOURCE: as Document 1, 'New Atlantis', 145–6, 156–7 and 162–5

31 Jonathan Swift
'GRAND ACADEMY OF LAGADO'

The Author permitted to see the Grand Academy of Lagado. The Academy largely described. The Arts wherein the professors employ themselves.

This Academy is not an entire single building, but a continuation of several houses on both sides of a street, which growing waste was purchased and applied to that use.

I was received very kindly by the Warden, and went for many days to the Academy. Every room hath in it one or more projectors, and I believe I could not be in fewer than five hundred rooms.

The first man I saw was of a meagre aspect, with sooty hands and face, his hair and beard long, ragged and singed in several places. His clothes, shirt and skin, were all of the same colour. He had been eight years upon a project for extracting sun-beams out of cucumbers, which were to be put into vials hermetically sealed, and let out to warm the air in raw inclement summers. He told me, he did not doubt, in eight years more, he should be able to supply the Governor's gardens with sunshine at a reasonable rate; but he complained that his stock was low, and entreated me to give him something as an encouragement to ingenuity, especially since this had been a very dear season for cucumbers. I made him a small present, for my lord had furnished me with money on purpose, because he knew their practice of begging from all who go to see them.

I went into another chamber, but was ready to hasten back, being almost overcome with a horrible stink. My conductor pressed me forward, conjuring me in a whisper to give no offence, which would be highly resented, and therefore I durst not so much as stop my nose. The projector of this cell was the most ancient student of the Academy; his face and beard were of a pale yellow; his hands and clothes daubed over with filth. When I was presented to him, he gave me a close embrace (a compliment I could well have excused). His employment from his first coming into the Academy, was an operation to reduce human excrement to its original food, by separating the several parts, removing the tincture which it receives from the gall, making the odour exhale, and scumming off the saliva. He had a weekly allowance from the society, of a vessel filled with human ordure, about the bigness of a Bristol barrel.

I saw another at work to calcine ice into gunpowder, who likewise showed me a treatise he had written concerning the malleability of fire, which he intended to publish.

There was a most ingenious architect who had contrived a new method for building houses, by beginning at the roof, and working downwards to the foundation, which he justified to me by the like practice of those two prudent insects, the bee and the spider.

There was a man born blind, who had several apprentices in his own condition: their employment was to mix colours for painters, which their master taught them to distinguish by feeling and smelling. It was indeed my misfortune to find them at that time not very perfect in their lessons, and the professor himself happened to be generally mistaken: this artist is much encouraged and esteemed by the whole fraternity.

In another apartment I was highly pleased with a projector, who had found a device for ploughing the ground with hogs, to save the charges of ploughs, cattle, and labour. The method is this: in an acre of ground you bury, at six inches distance and eight deep, a quantity of acorns, dates, chestnuts, and other mast or vegetables whereof these animals are fondest; then you drive six hundred or more of them into the field, where in a few days they will root up the whole ground in search of their food, and make it fit for sowing, at the same time manuring it with their dung. It is true, upon experiment they found the charge and trouble very great, and they had little or no crop. However, it is not doubted that this invention may be capable of great improvement.

I went into another room, where the walls and ceiling were all hung round with cobwebs, except a narrow passage for the artist to go in and out. At my entrance he called aloud to me not to disturb his webs. He lamented the fatal mistake the world had been so long in of using silk-worms, while we had such plenty of domestic insects, who infinitely excelled the former, because they understood how to weave as well as spin. And he proposed farther, that by employing spiders, the charge of dying silks should be wholly saved, whereof I was fully convinced when he showed me a vast number of flies most beautifully coloured, wherewith he fed his spiders, assuring us, that the webs would take a tincture from them; and as he had them in all hues, he hoped to fit everybody's fancy, as soon as he could find proper food for the flies, of certain gums, oils, and other glutinous matter to give a strength and consistence to the threads.

There was an astronomer who had undertaken to place a sun-dial upon the great weathercock on the town-house, by adjusting the annual and diurnal motions of the earth and sun, so as to answer and coincide with all accidental turnings by the wind.

I was complaining of a small fit of colic, upon which my conductor led me into a room, where a great physician resided, who was famous for curing that disease by contrary operations from the same instrument. He had a large pair of bellows with a long slender muzzle of ivory. This he conveyed eight inches up the anus, and drawing in the wind, he affirmed he could make the guts as lank as a dried bladder. But when the disease was more stubborn and violent, he let in the muzzle while the bellows were full of wind, which he discharged into the body of the patient, then withdrew the instrument to replenish it, clapping his thumb strongly against the orifice of the fundament; and this being repeated three or four times, the adventitious wind would rush out, bringing the noxious along with it (like water put into a pump), and the patient recover. I saw him try both experiments upon a dog, but could not discern any effect from the former. After the latter, the animal was ready to burst, and made so violent a discharge, as was very offensive to me and my companions. The dog died on the spot, and we left the doctor endeavouring to recover him by the same operation.

I visited many other apartments, but shall not trouble my reader with all the curiosities I observed, being studious of brevity.

I had hitherto seen only one side of the Academy, the other being appropriated to the advancers of speculative learning, of whom I shall say something when I have mentioned one illustrious person more, who is called among them *the universal artist*. He told us he had been thirty years employing his thoughts for the improvement of human life. He had two large rooms full of wonderful curiosities, and fifty men at work. Some were condensing air into a dry tangible substance, by extracting the nitre, and letting the aqueous or fluid particles percolate; others softening marble for pillows and pin-cushions; other petrifying the hoofs of a living horse to preserve them from foundering. The artist himself was at that time busy upon two great designs; the first, to sow land with chaff, wherein he affirmed the true seminal virtue to be contained, as he demonstrated by several experiments which I was not skilful enough to comprehend. The other was, by a certain composition of gums, minerals, and vegetables outwardly applied, to prevent the growth of wool upon two young lambs; and he hoped in a reasonable time to propagate the breed of naked sheep all over the kingdom.

SOURCE: Jonathan Swift. *Travels into Several Remote Nations of the World,* by Lemuel Gulliver, Part III, Chapter V (London, 1726)

L'ACADÉMIE DES SCIENCES

The French Academy of Sciences was sponsored by Colbert, Louis XIV's Minister, who was so concerned with economic and social life and its regulation in the latter half of the seventeenth century. The Academy arose from the ideas and meetings of Parisian men of letters and science in the mid-seventeenth century, who first proposed to emulate Richelieu's Académie Française, founded by Louis XIII's great Minister to foster the French language and encourage French linguistic and literary life.

The new scientific Academy soon collected an impressive band of men willing to research in the Bibliothèque Royale and willing to use some limited government funds to build observatories and finance less costly experiments.

The Academy was more secretive about its activities than was the English Royal Society; it released some details of its work from time to time in the Journal des Scavans, *while individual members published works of research, but it was not until the eighteenth century that a half century of research and enquiry could be collated, assessed and finally published. The results then released to the public displayed a greater stamina in pursuing any particular project than any other European Academy could muster, and reflected something of the self-conscious need for professionalism felt by the French body, which, as Justel tells us, had been very conscious in its early years of the need for achievement to justify its existence to Louis XIV.*

The first extract (32), taken from an English work of 1742 reviewing the philosophical and historical aspects of the Academy's life, tells something about the sources available for the early history of the Academy and gives some insight into its foundation. The second extract (33) takes from the English translation of the Memoirs of 1699 *some passages concerning the work performed during the course of that year by Fellows of the Academy. The two chosen show the Academy's interest in meteorology and in natural history. The final extract (34) provides a longer piece presented to the Society concerning a method for the throwing of bombs, an interest pursued by the Society, with obvious strategic implications of some value to the monarch who fostered the enterprise.*

32 John Martyn
'PHILOSOPHICAL AND HISTORICAL MEMOIRS': PREFACE

Soon after the foundation of the Royal Society, the late King of France, who was unwilling to omit any thing which might contribute to his glory, assembled a considerable number of the most able men in his kingdom, to whom he gave the title of the ROYAL ACADEMY OF SCIENCES; and engaged them in the pursuit of natural and mathematical knowledge. But this learned assembly was not established, as a regular society, till that monarch, in the beginning of the year 1699, thought fit to give them a new establishment, and to regulate their conduct by a body of laws framed for that purpose.

From the time of this new establishment of the Royal Academy of Sciences, a volume of their discoveries and observations has been published every year, under the title of *A History of the Royal Academy of Sciences*, by their perpetual secretary, the celebrated M. Fontenelle.

This *History* consists of two parts: the first is what is particularly called the *History*. In this part, M. Fontenelle gives a short account of every remarkable transaction, either in writing or by word of mouth, that has passed in the academy. The other part contains the *Memoirs*, or pieces that have been read in the academy, such as have been judged to be of the greatest importance, and the most worthy of being communicated to the publick at full length.

The papers contained in this curious *History* are ranged under several heads: the first of which comprehends those which relate to natural philosophy in general. Under this head the reader will find many remarkable accounts of the nature and properties of the air, observations on the thermometer, barometer, &c. discourses on the causes and effects of winds, on the flux and reflux of the sea, relations of the effects of thunder, lightning, &c. considerations on the aurora borealis, and other remarkable appearances in the heavens; curious discoveries of the various and surprising properties of the loadstone; inquiries into the nature of the magnetical needle; and many other entertaining as well as instructing particulars, which we shall not here detain him with enumerating.

To these papers on natural philosophy in general, we have added many others, which we thought would be esteemed to be the most useful and entertaining to the generality of readers. These contain the natural history and anatomy of animals, many useful discoveries relating to geography and navigation, a great variety of optical, dioptrical, and mechanical discoveries and observations, with many other mathematical papers, which relate to common uses of life, passing over those, which are merely speculative, or of less general use.

The Abridgment of the *Philosophical Transactions* has been so well received by the publick, that we flatter ourselves the present work, which is so nearly allied to it, will meet with a no less favourable reception. It is an Abridgment of all the *Memoirs of the Royal Academy of Sciences*, which relate to natural philosophy in general, to the natural history of animals, and to practical parts of the mathematicks; to which we have prefixed such accounts in M. Fontenelle's *History*, as are not to be found in the *Memoirs* themselves.

As for those accounts in the *History*, which are found more at large in the *Memoirs*, we have thought proper wholly to omit them; because they are generally no more than a historical account of what is contained in the *Memoirs* themselves. And as we have contracted each of those *Memoirs*, into as narrow a compass as we could, without injuring their sense, we thought it would be unnecessary to add the historical account of them too, which would swell the bulk of these volumes too much; it being our design to lay before the reader the philosophical discoveries and observations of the Royal Academy of Sciences, with relation to the particulars above mentioned, in as concise a manner as we could, without diminishing any part which really conduces to that end.

If the present undertaking meets with the success which we hope for, it will be an encouragement to us to proceed in abridging the other papers, which, tho' they are of less general use than these now published, will, however, be of the greatest service to such as are engaged in those particular studies.

SOURCE: John Martyn (trans). *The Philosophical History and Memoirs of the Royal Academy of Sciences at Paris*, Vol. I (London, 1742), iv–viii

I. *Comparisons of the observations made in different places, on the barometer, the winds, and quantity of rain.*

M. MARALDI having seen the observations made by Dr. Derham on the barometer and on the winds at Upminster, in England, during the years 1697 and 1698, compared them with those which were made at the observatory during the same years; and this is the result of the comparison.

Tho' the winds which reign at Paris, and at Upminster, are commonly different, yet there are many days in the different seasons of the year, in which the winds have been the same in both these places. When the wind has been the same in both, it has usually been violent, and of some continuance; tho' even these have sometimes varied.

There is found also some conformity in the constitution of the air, and it hath often happened to rain, or snow, or to be fair weather in both these places on the same days.

There is a great agreement between the variation of the height of the barometer observed at Paris and at Upminster. We generally find it to rise or fall at Paris, when it rises or falls at Upminster; tho' these variations are not always equal. In each month the days on which the quicksilver has been highest or lowest, have been the same both at Paris and Upminster; but commonly when it has been lowest, it has been 3 or 4 lines lower at Paris than at Upminster, the English measure being reduced to that of Paris.

It appears by the observations:

1. That the quicksilver rises sometimes, when the wind is N.N.E. or N.W. and that it falls with a S.S.E. and S.W. wind. It has however risen, and fallen at the same time in both these places, tho' the winds have often been different, and sometimes even opposite.

2. In these two last years, when the quicksilver was at the

lowest in both places, there fell some snow; it sometimes also snowed, tho' the quicksilver did not fall more than usual.

3. When the quicksilver rose, it was often fair weather, and it fell when it was rainy weather; it was often fair weather also when the quicksilver was low, and cloudy when it was high.

4. When the quicksilver fell at the same time in both places, and it rained in one and was fair in the other, the quicksilver often fell in proportion where it rained. In like manner when it rose in both places, it rose higher in proportion where it was fair.

Lastly, it appears, that the quicksilver rose in one when it fell in the other, and that it fell in like manner, whether the wind and weather were the same in both places, or different.

M. de Vauban having sent the academy a memoir on the quantity of rain that fell in the citadel of Lille for 10 years, from 1685 to 1694, M. de la Hire has compared the 6 last years of the observation at Lille with the same years, which he has observed very exactly at Paris.

Years.	At Lille.		At Paris.	
	Inches.	Lines.	Inches.	Lines.
1689	18	9	18	$11\frac{1}{2}$
1690	24	$8\frac{1}{2}$	23	$3\frac{3}{4}$
1691	15	2	14	$5\frac{1}{4}$
1692	25	$4\frac{1}{2}$	22	$7\frac{1}{2}$
1693	30	$3\frac{1}{2}$	22	8
1694	19	3	19	9
6 years	133	$6\frac{1}{2}$	121	9

By the comparison of these 6 years we see in general, that it rains rather more at Lille than at Paris; and that the mean year at Lille will be 22 inches, 3 lines, and at Paris 20 inches, 3 lines $\frac{1}{2}$.

But M. de la Hire has found in 1695, 19 inches, 7 lines $\frac{3}{4}$; in 1696, 19 inches, 5 lines $\frac{1}{2}$; in 1697, 20 inches, 3 lines; in 1698, 21 inches, 9 lines; and taking a mean year for these 10, we find 20 inches, 3 lines $\frac{1}{2}$ for each of the 6 first; whereas at Lille the

6 last give for a mean 22 inches, 3 lines, and the 10 together give 23 inches, 3 lines.

II. *Observations on the singularities of the natural history of* France.

The academy having a design to examine the wonders of France, has begun with Dauphiny, and with a very famous burning fountain in that province, four hours distant from Grenoble.

Saint Augustin has mention'd it, and seems to treat it as a supernatural miracle. But as it is good to be well assured of our facts, and not to seek for the reason of what does not exist; M. de la Hire wrote to M. Dieulamant, the king's engineer in the department of Grenoble, from whom he received an information, as well circumstanced as could be wished. M. Dieulamant went upon the spot, and examined it with a philosophical eye.

It is no fountain; but a small spot of ground, 6 feet long, and 3 or four broad, where may be seen a light, wandering flame, like that of brandy, fixt to a dead rock, of a sort of rotten slate, which crumbles in the air. This spot is upon a pretty steep declivity; about 12 feet below it, and as much on one side, there falls from the neighbouring mountains a little rivulet or torrent, which perhaps may formerly have flowed more high, and near the burning earth, which might have given room to think that the water burned.

The flame is not observed to come out of any hole or cleft in the rock, by which it might be suspected to have a communication with some lower cavern that might be inflamed. There is no matter found which may serve as nourishment to the flame, and it is only perceived to have a strong smell of sulphur; and it leaves no ashes. There is a very hot, white sort of salt petre about the place where the fire appears.

M. Dieulamant has been assured, that this fire burns most in winter, and in moist weather, that it diminishes gradually in great heats, and often goes out near the end of summer, after which it rekindles of itself. It is very easy also to make it kindle by another fire, which it does readily with a noise.

M. Dieulamant observed in the last place, that about the fire the earth cleaves, gives way, and sinks down. He does not ascribe this however to the fire, but to the waters which flow among the dead rocks, and hollow the earth, or carry it away. This effect is so considerable in Dauphiny, and especially in the country called le Chanseaux, that sometimes two villages situated upon two different mountains, which could not be seen from each other, because other higher mountains were between them, have begun at once to be seen from the sinking of the interposed mountains.

SOURCE: as Document 32, 11–15

34 L'Académie des Sciences
'MEMOIRS, 1700'

V. *A general method of throwing bombs in all cases proposed, with an universal instrument for this purpose, by* M. de la Hire.*

This† instrument consists only of a semicircular plate ABD, which has a rule or tail BE fixed upon the edge of the circle, which answers to the centre C, and of which the side BE being prolonged, meets the centre C of the circle, and is perpendicular to the diameter ACD of the semicircle. The semidiameter CD of the semicircle is divided into 9 equal parts, and upon CE there are also marked the same divisions, as upon CD. Each of these divisions represents 100 toises, and they may be subdivided by little points into other lesser parts; but it will be easy to judge by the eye alone, which is sufficient for these sorts of operations. Through the divisions of the diameter 7, 8 and 9, draw semicircles with their centre in C.

To the centre C of the instrument, is fixed a rule CF, which is slit, in the middle, that it may let the pin G run in it, with the head turned toward the plate; the middle of this slit answers to the centre C of the semicircle. This pin G passes also into the slit of another rule IH like the first, so that these two

* July 24, 1700.   † Fig. 6 [see p 165].

rules may be held fast by the same pin, with a little screw with ears, to what length, and in what angle you please.

At the extremity H of the rule IH, there is a little round plate, the breadth of the rule, which should be of some white metal, as silver, upon which is marked a black point, which answers to the middle of the slit of the rule. Upon this rule IH, from the centre H of the silver pin are also marked the magnitudes C7, C8, C9.

The rule CF, which may turn freely upon its pin C, which is placed in the centre of the semicircle, may be held fast in what position you please, upon the plate of the semicircle, by the means of a screw at the end of the pin, which locks the plate and rule together, it must be observed, that these pins C and G must be pointed at their ends, and raised a little above their screw.

Along the semidiameter CD, there is upon the plate a slit, into which passes or runs a sliding piece O, with a pointed head, to which is fastened the thread of a plummet: the middle of this slit must be upon the diameter ACD of the semicircle.

As for the use of this instrument, it may serve at first to determine the distance from the station where the mortar is, to the mark, without its being necessary to know the number of toises of this distance. This is done by taking some base of 3 or 4 hundred toises on the side of the place where one is, and aiming at some known object at this distance, by the pin C and the sliding piece G; aim also at the mark by the pins C and G, fix the rule CF, in this position on the plate, by means of the screw at C. Afterwards remove the instrument to the station where you aimed at first, and then place the sliding piece O at the number of toises from the point C to the other station: and aiming also by the sliding piece and by the pin C, make the pin G run upon the rule CG till you see by the sliding piece O and the pin G, the mark you aim at: then stop the pin G in this position, and the distance CG will be in the parts of CO the number of toises from the station where you shoot, to the mark; which is not necessary to know, provided the pin G is stopped upon the rule CF in this position. But if this distance was given, you need only place the rule CF upon the line CE, and stop the pin G in the number of toises there marked.

To have the elevation or depression of the mark, with relation to the horizon of the place from whence you are to shoot, the plate must be held fast in such a situation, that the thread of the plummet P be applied to the point A, or upon the diameter AD, and move the rule CF, so that you may see the mark by the point of the pins C, G. You must then stop the rule CF very fast upon the plate of the semicircle. These are the two things we must always know, together with the force of the powder in whatsoever method it be.

As for the force of any certain quantity of powder, or of the height to which the cast can be raised in aiming toward the zenith; as it cannot be known by experiment, we know by the demonstration that the point of the horizon where the bomb may come, when it is cast in an angle of 45°, or half a right one, is always distant from the place where you aim, double the elevation of the vertical cast; this is what is called the amplitude of the cast; wherefore one single observation made in this manner, and with its conveniencies, may serve for all sorts of cases. For example, if the cast is 1600 toises, the vertical or perpendicular cast will be raised to 800 toises. This instrument may also serve to make this cast, as will be seen hereafter.

Now for the practice of casts, and the use of the instrument, you must place the sliding piece O upon the number of toises of the vertical cast, as in the figure, upon the point of 900 toises, if the vertical cast is 900 toises, or else if the amplitude of the cast be 45°, 1800 toises. Afterwards move the rule IH in its groove upon the pin G immoveable upon CF, till the same division of 900 toises of this rule HI meet upon CE, when IH will be pretty near parallel to AD, or IH perpendicular to CE; which will be known by turning the rule IH, if its division of 900 touches the line CE. Then if you turn the rule IH upon the pin G, the length GH remaining always the same, that you found it. When the centre of the plate H shall touch the circle OLL of 900 toises, stop the rule IH very fast to the rule CF with the screw which is at G. Now if you apply one of the sides of the tail BE, which we suppose of equal breadth everywhere, within the mortar or cannon, and afterwards raise or depress it as much as is necessary to make the thread of the plummet P pass by the point L, the cast will fall at the mark designed.

If the point H cannot meet its circle but in one point, there is but one single cast, which can go to the mark; but if it can meet it in 2 points, these 2 points, as L, will serve to make 2 different casts in the same manner, that we have explained for one, which will both go to the same mark. If the point H cannot meet its circle, there is not any cast which can go to the mark.

The demonstration

* The demonstration of this practice depends upon a proposition of the parabola, which I have demonstrated, and which M. Blondel relates from me in Chap. 7. of part III. of his treatise. For the point G being the mark aimed at, and by my construction GH being equal to the difference, between the height of the vertical cast CO, and the height of the mark G above the horizon, which passes by C, or else the sum of the same, if the point G is below the horizon; or under the level of the point C, that is, if GH is perpendicular to the horizon, the line OH will be the level: therefore the points of intersection LL of the two circles OL, HL, are the foci of the parabola's, which will pass by C and G, and it is this figure which M. Blondel has only reversed in his square instrument, which does not alter the construction.

Now I say that the line OL being vertical or perpendicular, will make with OH or CE of my instrument, which is parallel to it, the angle that the direction of the cast must make to the point C, without there being need to make a new operation, and it is in this chiefly that the simplicity of this practice consists.

For if you draw LM perpendicular upon OH, and bisect it in S, the point S must be the summit of the parabola of the cast. But also by the properties of the parabola, we know that the line LC being drawn, and OC being a diameter, the tangent CN of the parabola at the point C, will bisect the angle OCL,

* Fig. 7 [see p 165].

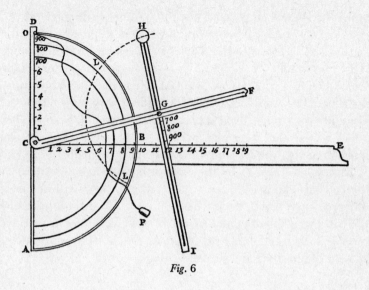

Fig. 6

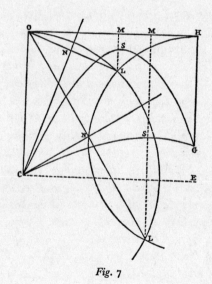

Fig. 7

and consequently the angle OCN determines the angle of the cast, if OC is considered as vertical. But I say also that the angle HOL is equal to the angle OCN, which is evident, since the 2 rectangular triangles CON, OLM, have one equal angle CON, OLM, the two lines CO, LM, being parallel, and cut by the same OL: therefore in the instrument, if the thread of the plummet passes by OL, it is plain that OH or CE, or any other which is parallel to them, will give the inclination of the mortar, or the line by which the cast must be made.

It may be observed by the casts of bombs or bullets, that as there are in almost all casts 2 points as LL, there will be also 2 different casts which go to the same mark, but the upper one will be more fit, to make its effort against bodies which are placed in a level, and the other cast will have more effect against bodies that are perpendicular, as walls.

SOURCE: as Document 32, 237–42

ROYAL SOCIETY

The sources for the Royal Society's activities have been preserved in their archives, and can be found in more accessible form in Birch's History, *the series of the* Philosophical Transactions, *Sprat's* History *and defence of the early days of the institution, and in the records of the individual members to be found scattered in archives and printed sources across the country. Founded from a nucleus of men who had pursued studies in natural philosophy in Oxford and London during the period of the Civil Wars — men of the stamp of Wallis, Wilkins and Ward — it soon became a fashionable London society patronised by men like Brouncker and the King himself. It received a royal charter, and often found itself blessed or burdened with projects or specimens sent down by the King, who preserved a dilettante interest in the Society and its activities throughout the early years of his reign.*

The materials reproduced below show the range of activities of the Society, and give, through its historians, something of the flavour of its deliberations and problems in its early years.

35 Thomas Birch
'EARLY RULES OF THE ROYAL SOCIETY (1660)'

ROYAL SOCIETY OF LONDON

That it be referred to the lord viscount BROUNCKER, Mr. BOYLE, Sir ROBERT MORAY, Dr. PETTY, and Mr. WREN, to prepare some questions, in order to the tryal of the quicksilver experiment upon Teneriffe.

It was likewise agreed, that the number of members should not be increased, but by consent of the society, who had this day subscribed their names, till the orders for the constitution should be settled.

The form of subscribing was in the following words:

'We, whose names are underwritten, do consent and agree, that we will meet together weekly, (if not hindered by necessary occasions) to consult and debate concerning the promoting of experimental learning: and that each of us will allow one shilling weekly towards the defraying of occasional charges: provided, that if any one, or more, of us shall think fit at any time to withdraw, he, or they, shall, after notice thereof given to the company at a meeting, be freed from this obligation for the future.'

It was farther agreed, that any three or more of the company, whose occasions would permit them, should be desired to meet as a committee at three of the clock on the Friday following, to consult about such orders, in reference to the constitution of the society, as they should think fit to offer to the whole company; and so to adjourn from day to day.

The society meeting again on the 12th of December, it was referred to the lord viscount BROUNCKER, Sir ROBERT MORAY, Sir PAUL NEILE, Mr. MATTHEW WREN, Dr. GODDARD, and Mr. CHRISTOPHER WREN, to consult about a convenient place for the weekly meeting of the society.

It was then voted, that no person should be admitted into

the society, without scrutiny, except such as were of, or above, the degree of baron:

That the stated number of the society be fifty five:

That twenty one of the said number be the *quorum* for elections:

That any person of, or above, the degree of baron might be admitted as supernumeraries, if they should desire it, and would conform themselves to such orders, as were or should be established.

And it having been suggested at the committee appointed at the preceding meeting, that the college of physicians would afford convenient accommodation for the assemblies of the society, upon supposition, that it were granted and accepted of, it was thought reasonable, that any of the fellows of the said college, if they should desire it, be admitted likewise as supernumeraries, upon condition of submitting to the laws of the society, both as to the payment on their admission and the weekly allowance, and the particular works or talks, that should be allotted to them.

It was also agreed, that the public professors of mathematics, physic, and natural philosophy of both universities, should have the same privilege with the college of physicians, on the same condition of paying the admission fee, and contributing their weekly allowance and assistance, when their occasions permit them to be in London.

The following regulations were likewise resolved upon:

That the quorum of the society be nine for all matters, except the business of election.

Concerning the manner of elections:

That no person be elected the same day, on which he is proposed.

That at least twenty one members be present at each election.

That the amanuensis provide several little scrolls of paper of an equal length and breadth, in number double to the members present. One half of these to be marked with a cross, and the other with cyphers; and both being rolled up, to be laid in two distinct heaps. Every person then coming in his order shall take from each heap a roll, and throw which he shall please privately into an urn, and the other into a box. After which, the director and two others of the society, having openly numbered the crossed rolls in the urn, shall accordingly pronounce the election.

That if two thirds of the members present do consent upon any scrutiny, that election be good, and not otherwise.

Concerning the officers and servants of the society:
That the standing officers of the society be three, a president or director, a treasurer, and a register:
That the president be chosen monthly.
That the treasurer continue one year; as also the register.
That there be likewise two servants belonging to the society, an amanuensis, and an operator.
That the treasurer give in every quarter an account of the stock in his hand, and all disbursements made, to the president or director, and any three others to be appointed by the society; who are to report this account to the society.

SOURCE: Thomas Birch. *The History of the Royal Society*, 4 vols (London, 1756–7), 5–6

36 Thomas Birch
'A MEETING OF THE ROYAL SOCIETY, 8 OCTOBER 1662'

Mr. JAMES HAYES was admitted a member of the society.
Mr. PETTY proposed WILLIAM HOARE, M.D. as a candidate.
Dr. TERNE was put to the scrutiny, and unanimously elected.
Dr. GODDARD and Dr. WILKINS were appointed curators, for the experiment proposed at the last meeting, how long a lamp will continue to burn under water.
Mr. WILLUGHBY produced his demonstration to prove, that the same area of ground planted with trees after a quincuncial

figure, will hold more trees placed at the same distance from one another, than the square, in the proportion of 8 to 7.

It was ordered, that the thanks of the society be given to Dr. MERRET, for his pains in translating the Italian discourse *De Arte Vitrariâ*, upon the motion and desire of the society: And

That Mr. WILDE be desired to communicate to the society such choice discourses, as he hath collected upon several subjects relating to experimental philsophy, that they may be perused and considered by several members of the society to be appointed for that purpose.

Mr. BOYLE was desired to shew, at the next meeting, the second part of his experiment about coagulation, viz. the reducing the coagulated liquors to their former fluidity.

It was debated and put to the question, whether the former order concerning admitting all persons of the degree of barons or above it, and all of his majesty's privy-council, upon their desire and subscription, without putting them to the scrutiny, should be continued and made a statute; and it passed in the affirmative, twenty one members and more being present.

Dr. WREN offered an experiment about the undulation of quicksilver in a crooked tube; which, he suggested, was, for the velocity of it, proportionable to the vibrations of a pendulum. He was desired to prosecute that experiment farther, and to give in an account of it.

The experiment of breaking wires was deferred till the next meeting.

Mr. OLDENBURG produced a letter, dated at Zurich, 17 Sept. 1662, giving an account of a new mixture of metals very useful for pistols and guns, highly esteemed in Germany, extremely light, and not subject to rusting or breaking. The following extract of this letter was ordered to be entered in the letter book.

'Here is a goldsmith, called FELIX WARDER, citizen of Zurich, who hath the invention of a way to make hand-guns and pistols, of a metal, which is called *orichalcum*. None besides himself is known to have this invention; and he tells me, that at Nuremberg they have spent thousands in making experiments to find it out, but all in vain. He can make that brittle metal so tough, that it is a great deal freer from the danger of breaking than iron: besides, the pistols and guns made of it are so thin and so light, that they weigh scarcely half the weight of iron ones, and will bear a double charge, without ever breaking or cracking. The grandees have been presented with these guns frequently, and he sells the least of these pistols for six doublons a pair; the larger sort, being gilt, for twelve doublons or Spanish pistoles. I have discoursed with him about the preparing this metal: all I can learn of him is this, that he adds something to the *orichalcum*, which maketh it fusible and tough. And the barrels of his pistols and guns are all cast, and then prepared for use, when they are scowered within and without. I did ask him, whether he was not willing to teach the art to any? He said he was, though he had taught it nobody yet but his daughter, now dead, who was wont to help him to make them, when he had so much work, that he could not dispatch it all alone. I enquired further, what his price would be for teaching it? He said, if any, to whom he should teach, it would bind him to teach none else, his price would be 200 doublons; but if it were left free to him to teach others, he would sell it for half that money. He added, that the invention could not be taught in writing, but that such, as would purchase it, must be present to see all themselves what is done, because that, besides the ingredients and the ordering of them, there is a certain slight, which if not seen and well observed, the work will miscarry.'

SOURCE: as Document 35, 115–16

ACCADEMIA DEL CIMENTO

The following short extract (37) illustrates something of the operation of one of the smaller European societies, at a time when they were just beginning to proliferate. By the end of the century everywhere from Dublin to Leipzig would have its equivalent, but the idea had been pioneered half a century earlier in Italy, developing out of the neo-platonic Florentine academies, and institutions like the Accademia dei Lincei.

The extract is taken from Waller's translation, published in England to bring to English audiences an idea of the sophistication of Italian glass-ware and experimentation; it includes a record of an experiment performed by the Accademia in the field of magnetism.

37 Accademia del Cimento
'NATURAL EXPERIMENTS'

DEDICATIONS

To Sir John Hoskyns Knight and Baronet, President of the Royal Society, etc.

Sir,

As your Commands gave the first being to this Attempt, so 'tis but Justice to offer it to your Self; and 'twas but necessary to crave so advantagious a Protection, to defend it against the Difficulties, things of this Nature meet with, in this Censorious Age.

I shall wave, as less grateful to you, a large Description of the Happiness the Royal Society enjoys under such a President, whose perspicacious Judgment is actuated by a true desire of promoting real Knowledge; and shall rather give some account of the Work it self: It was presented in a Publique Meeting of the Royal Society, March 12. 1667/8 by Sigr. Lorenzo Maga-lotti, and Sigr. Paulo Falconieri, from the Most Serene Prince Leopold, Brother to Ferdinand the Second, Great Duke of Tuscany; and has ever since layn in our Library expecting a more skilful Pen, to perform what I have here aimed at. The Experiments are many, and curious, made under the favour of that Prince, by the Members of the Academy Del Cimento, men of great ingenuity; and related with much sincerity by the Secretary of that Academy; which Society (I hear) is now scatter'd, and the Hopes of those Benefits the Learned World might justly expect from them, frustrated. Many indeed of these Experiments have been made, and shewn in several Meetings of the Royal Society (before, and since the Publication of this in the Italian, in the Year 1667) by the Honourable Robert Boyle, Esq; and other worthy Members thereof; but for all this, I hope it may not prove unacceptable to find the Ingenious in other Parts of the World, have not thought their time mispent in these Endeavours, what contrary Sentiments soever some may have; nor will the agreement between the success of Experiments made there, and what has been attempted here (often with a differing Apparatus) be less pleasing: very many, I dare undertake, are New to most Persons, except your self, and upon that account will prove more diverting. I need not add the great Expence of Care, and Charge, and Fatigues of the Academy in this Work; nor the scarcity of this Piece in the Original, no small Motive to this Undertaking (that it might be obtained with more Ease, and at a cheaper Rate;) which how performed, I submit to your Self, and the worthy Members of the most Illustrious Royal Society . . .

To the Most Serene
Ferdinand II.
Grand Duke of Tuscany.

Most Serene Prince,
The Publishing of these First Essays of *Natural Experiments*, which for many Years have been made in our Academy, under the Protection, and with the Indefatigable Assistance of the most Serene Prince Leopold your Highness's Brother, will prove the happy occasion of giving fresh Testimonies (of your Highness's great Liberality) to all those parts of the World where Vertue is adorn'd with its deserved Lustre; and will create a new sense of Gratitude and Respect in all true Lovers of the more curious Arts, and Nobler Sciences. Especially we ought to frame our thoughts to a more humble Acknowledgment, as we are more nearly concerned and warmed by the cherishing Rays, and invigorating Influence of your Highness's bounty. Which with the favour of your Patronage, the incourag-

ing invitation of your Mind, and proper Genius that way; but above all, with the Honour of your Presence, sometimes stooping to our Academy, sometimes commanding us to your Royal Apartments, has bestowed upon it an Immortal Name; Kindled Active Desires in our Breasts, and given an happy encrease to our Studies. These considerations easily demonstrate, with what duty we are engaged to Consecrate the first Fruits of our Labours to your Highness's most Illustrious Name; since nothing can proceed from us, wherein you can have a greater share, and by consequence more due to you; nor any thing that may make fairer approaches to merit the happy Fate of your generous Acceptance. 'Tis certain, that through the Excess of so large and signal Favours, we can be sensible of no greater Resentments than to find our selves so much obliged to your Highness: not that we refuse to bear the Weight of so endearing and inestimable an Obligation; but onely because we would wish to be able to offer something not purely your own; whence we might at least flatter our selves, That we had made some small return which your Highness might impute in some degree to our choice, and not wholy redewable to your Highness Self, or Necessity. But we must rest satisfied with the bare desire of so just and deserved a Passion; since these new Philosophical Speculations are so deeply Radicated in your Highness's Protection, that not onely what is now produced by our Academy, but what ever shall be brought to Maturity in the most Famous Schools of Europe, or After Ages raise up, shall be likewise due to your Highness, as the gift of your Beneficence: since as long as the Sun, Planets, and Stars retain their glory, and Heaven endures, there will remain a glorious Memory of one that contributes so much with his auspicious Influence to such new and strange Discoveries; opening an unbeaten Path for the least fallacious Method of search after Truth. Yet in so great a scarcity of Tributes, some little thing presents it self to manifest our grateful observance; which is the onely joy wherewith we support our Deficiency, while all redounds in more resplendant Glory to your Highness, who having already acted your full Proportion of what ever new, good, and great, is at

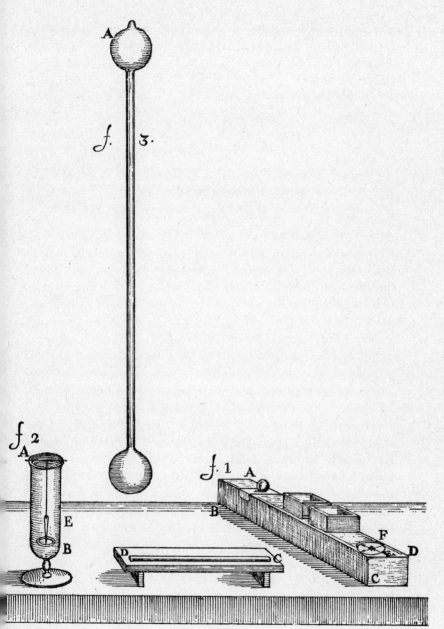

Tab. 19, *Fig.* 1

175

any time to be found in the Repository of Sciences, has ener-
vated and discouraged all thoughts of emulation in others.
This, and this alone are we able to lay at your Highness Feet,
whose continual Protection, we crave with Respect and
Reverence, begging from Heaven the height of Prosperity and
Grandure to your Highness.

THE FIRST EXPERIMENT

To discover if (except Iron *or* Steel) *there be any* Solid, *or* Fluid
Body, which interposed between the Magnet *and* Iron, *will cause any*
variation, *or quite cut off the Passage* to the Magnetic *Virtue.*

At one end of a Wooden Box ABCD [Tab. 19, Fig. 1], we fixt a
Copmass, and opposite to the Dart (respecting the point E) at
the other end of the Box, we moved a Magnet, and gently
approaching it nearer till the Dart was removed one Degree,
that is, from E, to F, we there fixt the Load-Stone, and in the
space remaining between it, and the Compass, we set either
glasses filled with Mercury, or Wooden Vessels filled with
Sand, or Fileings of Metal, (except of Iron, or Steel) or solid
Parallelipipid's of the same Metal, or of divers Stones, and
Marbles; but still we found the Dart unmoved from the Point
F. Lastly, we filled the same Vessels with Spirit of Wine, and
set it on fire, yet that Flame did not in the least divert the
Power attracting the Dart to F: and a thin Plate of Iron or
Steel, onely was able to vary it, and make it return to E, as is
already known. And not onely the above-named Causes were
unable to obstruct the Magnetic Activity; but we have laid
upon one another 50 pieces of Gold, and laid a Needle upon
the uppermost, which has obeyed the Motion of a Magnet,
moved under the lowermost.

SOURCE: R. Waller (trans). *Essayes of Natural Experiments Made
in the Academie del Cimento* (London, 1684), Dedications and 124

Part Four

INSTRUMENTS AND MEN

The seventeenth-century men of science used their own skills and imagination in conjunction with the skills of artisans and technologists to produce new instruments capable of furthering scientific enquiry. The best known and the most exciting to their contemporaries were the telescope and the microscope, carefully harnessing the power of dioptrics to extend the range of the senses. But these were by no means the only productions of this century of experiment. In addition there were Boyle's air pump, the Toricellian mercury column that was to become a barometer, the Magdeburg hemispheres capable of demonstrating the force of air pressure upon a vacuum, improved surveying instruments, the mariner's compass, improved charts, and clocks developed way beyond the standards of sixteenth-century timepieces. One has only to look at most seventeenth-century instruments to notice the way in which men were pursuing accuracy with more and more vigour; they wished to measure smaller units of time, survey with finer divisions of minutes and degrees, chart with greater accuracy, measure the circumference of the world with new precision, gauge forces in the physical universe with greater accuracy and extend their senses by whatever aids they could engineer.

The extracts beneath concentrate upon the advances made in dioptrics. Power put forward extravagant claims for the further development of the microscope in his preface to his microscopical observations (38), and Hooke's observations of a Flea (39) illustrate the kind of results that men could obtain from these new-found aids to science. Both the telescope and especially the microscope rapidly made their way into the average gentleman's household and became toys and playthings; but at the same time the microscope and telescope did have serious scientific implications when built to high standards and developed to a greater capacity by the later seventeenth-century craftsmen.

The men that made the movement are better documented in England than elsewhere. The English vogue for diaries and observations, for gossip and chit-chat has left us with an abundance of contemporary material of a biographical kind. Aubrey's Brief Lives, Pepys's Diary, and Ward's Lives of the Gresham Professors are indispensable sources. Reproduced below are Ward's accounts of Gunter and Gellibrand, both sometime Gresham professors (40), and Aubrey's comments on Thomas Hariot and Edmund Halley (41).

THE MICROSCOPE

38 Henry Power
'SOME OBSERVATIONS WITH THE MICROSCOPE'

OBSERVAT. XXXIII.

Corns of Sand, Sugar, and Salt

It is worth an Hour-glass of Time to behold the Crystal Sands that measure it; for they all seem like Fragments of Crystal, or Alum, perfectly Tralucent, of irregular polyhedrical figures, not any one globular; every Corn about the bigness of a Nuttmeg, or a Walnutt: which from their unequal superficies refracting and reflecting the Sun rays, seem here and there of Rainbow-colours.

Being layd of a row or train, they seemed like a Cawsy of Crystal Stones, or pure Alum Lumps: So that now we need not so much wonder with the Vulgar Philosophers, how so clear and glorious a body as glass, should be made of so durty, opace, and contemptible Materials, as Ashes and Sand; since now we are taught by this Observation that Sand, and Salt which is in the Ashes, the two prime Materials thereof, are of themselves so clear and transparent, before they unite into that diaphanous Composition.

A Small Atom of Quick-Silver

An Atom of quick-silver (no bigger than the smallest pins-head) seemed like a globular Looking-glass) where (as in a Mirrour) you might see all the circumambient Bodies; the very Stancheons and Panes in the Glass-windows, did most clearly and distinctly appear in it: and whereas, in most other Mettals, you may perceive holes, pores, and cavities; yet in ☿ none at all are discoverable; the smallest Atom whereof, and such an one, as was to the bare Eye, *tantùm non invisibile*, was presented as big as a Rounseval-Pea, and projecting a shade; Nay, two other Atoms of ☿, which were casually layd on the same plate, and were undiscernable to the bare eye, were fairly presented by our Microscope.

OBSERVAT. LI

Of Aromatical, Electrical, and Magnetical Effluxions

Some with a Magisterial Confidence do rant so high as to tell us, that there are Glasses, which will represent not onely the Aromatical and Electrical Effluxions of Bodies, but even the subtile effluviums of the Load-stone it self, whose Exspirations (saith Doctor Highmore) some by the help of Glasses have seen in the form of a Mist to flow from the Load-stone. This Experiment indeed would be an incomparable Eviction of the Corporeity of Magnetical Effluviums, and sensibly decide the Controversie 'twixt the Peripatetick and Atomical Philosophers.

But I am sure he had better Eyes, or else better Glasses, or both, then ever I saw, that performed so subtle an Experiment: For the best Glasses that ever I saw, would not represent to me, the evaporations of Camphire (which spends it self by continually effluviating its own Component Particles;) nay, I could never see the grosser steams that continually perspire out of our own Bodies, which you see will soil and besmear a polished Glass at any time; and which are the fuliginous Eructations of that internal fire, that constantly burns within us.

Indeed if our Dioptricks could attain to that curiosity as to grind us such Glasses, as would present the Effluviums of the Magnet, we might hazard at last the discovery of Spiritualities themselves: however it would be of incomparable use to our Modern Corpuscularian Philosophers, who have banished qualities out of the list of the Predicaments. And truly, as the Learned Doctor Brown hath it; The Doctrine of Effluxions, their penetrating Natures, their invisible paths, and unsuspected effects, are very considerable: for (besides the Magnetical One of the Earth) several Effusions there may be from divers other Bodies, which invisibly act their parts at any time, and perhaps through any Medium: A part of Philosophy but yet in discovery; and will, I fear, prove the last Leaf to be turned over in the Book of Nature.

Some Considerations, Corollaries, and Deductions, Anatomical, Physical, and Optical, drawn from the former Experiments and Observations.
First, Therefore, it is Ocularly manifest from the former Observations, that, as perfect Animals have an incessant motion of their Heart, and Circulation of their Bloud (first discovered by the illustrious Doctor Harvey;) so in these puny automata, and exsanguineous pieces of Nature, there is the same pulsing Organ, and Circulation of their Nutritive Humour also: as is demonstrated by OBSERV. fourth, sixth, seventeenth, etc.

Nay, by OBSERV. sixth, it is plain that a Louse is a Sanguineous Animal, and hath both an Heart and Auricles, the one manifestly preceding the pulse of the other; and hath a purple Liquor or Bloud, which circulates in her (as the Noblest sort of Animals have) which though it be onely conspicuous in its greatest bulk, at the heart, yet certainly it is carried up and down in Circulatory Vessels; which Veins and Arteries are so exceeding little, that both they and their Liquor are insensible: For certainly, if we can at a Lamp-Furnace draw out such small Capillary Pipes of Glass that the reddest Liquor in the World shall not be seen in them (which I have often tried and done;) how much more curiously can Nature weave the Vessels of the Body; nay, and bore them too with such a Drill, as the Art of man cannot excogitate: Besides, we see, even in our own Eyes, that the Sanguineous Vessels that run along the white of the eye (nay and probably into the diaphanous humours also) are not discernable, but when they are preter-naturally distended in an Ophthalmia, and so grow turgent and conspicuous.

To which we may adde, that in most quick Fish, though you cut a piece of their flesh off, yet will no bloud be discernable, though they be sanguineous Animals; but the bloud is so divided by the minuteness of their Capillary Vessels, or percribration through the habit of the Parts, that either it has lost its redness, or our eyes are not able to discover its tincture.

Secondly, It is observable also from the former Experiments, that in these minute Animals their nutritive Liquor never arises to the perfection of bloud, but continually as it were remains Chyle within them, for want of a higher heat to dye it into that Spirituous Liquor: Nay, you shall observe in perfect Sanquineous Animals a Circulation of an albugineous chylie-matter (before the bloud have a being) if you take Nature at the rise, and critically observe her in her rudimental and obscure beginnings.

For view but an Egge, after the second day's Incubation, and you shall see the cicatricula in the Yolk, dilated to the breadth of a groat or six-pence into transparent concentrical circles; in the Centre whereof is a white Spot, with small white threads, (which in futurity proves the Heart with its Veins and arteries) but at present both its motion and circulation is undiscernable to the bare eye, by reason of the feebleness thereof, and also because both the Liquor and its Vessels were concolour to the white of the Eggs they swum in; but the Heart does circulate this serous diaphanous Liquor; before (by a higher heat) it be turned into bloud.

And one thing here I am tempted to annex, which is a pretty and beneficial Observation of the Microscope, and that is, That as soon as ever you can see this red pulsing Particle appear (which Doctor Harvey conceived, not to be the Heart, but one of its Auricles) you shall most distinctly see it, to be the whole Heart with both Auricles and both Ventricles, the one manifestly preceding the pulse of the other (which two motions the bare eye judges to be Synchronical) and without any interloping perisystole at all: So admirable is every Organ of this Machine of ours framed, that every part within us is intirely made, when the whole Organ seems too little to have any parts at all.

Thirdly, It is peculiarly remarkable from Observation xxxi. That not onely the bloud in perfect Animals, and the chyle in imperfect ones; but also the Animal Spirits have a Circulation, which singular observation hath often provoked and entised our endeavours into a further enquiry after the Nature of these Spirits, as to their Origin or Generation, their activity and motion, with some other eminent properties belonging to them: we shall draw our thoughts together, and so present them to your View: I will not say, that our discourse hereon, shall pass for an uncontrollable authentick Truth; it is all my ambition if it attain but to the favourable reception of a rational Hypothesis at last.

SOURCE: Henry Power. *Experimental Philosophy* (London, 1664), 42–3 and 57–61

39 Robert Hooke
'MICROGRAPHIA'

Observ. LIII. Of a Flea.

The strength and beauty of this small creature, had it no other relation at all to man, would deserve a description.

For its strength, the Microscope is able to make no greater discoveries of it then the naked eye, but onely the curious contrivance of its leggs and joints, for the exerting that strength, is very plainly manifested, such as no other creature, I have yet observ'd, has any thing like it; for the joints of it are so adapted, that he can, as 'twere, fold them short one within another; and suddenly stretch, or spring them out to their whole length, that is, of the fore-leggs, the part A, of the 34. scheme, lies within B, and B within C, parallel to, or side by side each other; but the parts of the two next, lie quite contrary, that is, D without E, and E without F, but parallel also; but the parts of the hinder legges, G, H and I, bend one within another, like the parts of a double jointed Ruler, or like the foot, legg and thigh of a man; these six leggs he clitches up altogether, and when he leaps, springs them all out, and thereby exerts his whole strength at once.

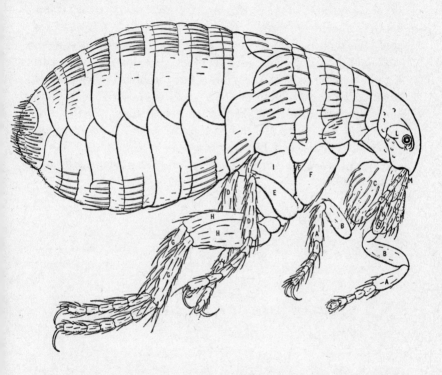

But, as for the beauty of it, the Microscope manifests it to be all over adorn'd with a curiously polish'd suit of sable Armour, neatly jointed, and beset with multitudes of sharp pinns, shap'd almost like Porcupine's Quills, or bright conical Steel-bodkins; the head is on either side beautify'd with a quick and round black eye K, behind each of which also appears a small cavity, L, in which he seems to move to and fro a certain thin film beset with many small transparent hairs, which probably may be his ears; in the forepart of his head, between the two fore-leggs, he has two small long jointed feelers, or rather smellers, MM, which have four joints, and are hairy, like those of several other creatures; between these, it has a small proboscis, or probe, NNO, that seems to consist of a tube NN, and a tongue or sucker O, which I have perceiv'd him to flip in and out. Besides these, it has also two chaps or biters PP, which are somewhat like those of an Ant, but I could not perceive them tooth'd; these were shap'd very like the blades of a pair of round top'd Scizers, and were opened and shut just after the same manner; with these Instruments does this little busie Creature bite and pierce the skin, and suck out the blood of an animal, leaving the skin inflamed with a small round red spot. These parts are very difficult to be discovered, because, for the most part, they lye covered between the fore-legs. There are many other particulars, which, being more obvious, and affording no great matter of information, I shall pass by, and refer the Reader to the Figure.

SOURCE: as Document 9, 210–11

BIOGRAPHIES

40 John Ward
'LIVES OF THE GRESHAM PROFESSORS'

III.

EDMUND GUNTER was born in the county of Hertford, but descended originally from Gunterstown in Brecknockshire. He was educated on the royal foundation at Westminster School, and elected from thence to Christ Church college in Oxford in the year 1599, being then eighteen years of age, where he took the degrees in arts. Mathematics were the prevailing studies of his youth, and about the year 1606 he invented the sector, and wrote the description and use of it in Latin; many copies of which were taken in writing, but none of them printed. After this he took orders, became a preacher, in 1614 was admited to read the *Sentences,* and proceeded to the degree of batchelor in divinity.

But his genius still leading him cheifly to mathematical pursuits, when Mr. Williams resigned the professorship of astronomy in Gresham college, he was chosen to succeed him March the 6, 1619, two days after his surrender. I mention this to obviate a mistake of Mr. Oughtred in the following passage of his *Apologeticall epistle.* 'In the spring 1618 (sais he) I being at London went to see my honoured freind, master Henry Briggs, at Gresham college, who then brought me acquainted with master Gunter, lately chosen astronomie reader there, and was at that time in doctour Brooke his chamber; with whom falling into speech about his quadrant, I shewed him my *horizontall instrument.*' And presently after he mentions a letter, he received from Mr. Briggs, dated from Gresham college 2 June 1618, and a postscript 4 June, which came to his hands June 10; in which letter of Mr. Briggs were these words: 'Master Gunter doth here send you the print of a horizontall diall of his drawing after your instrument.' It is plain from the account given above concerning Mr. Gunter's election, that Mr. Oughtred was mistaken in saying, he was chosen astronomy lecturer in Gresham college in 1618. But as his *Apologeticall epistle* was not written till many years after Mr. Gunter's death, a slip of his memory in such a circumstance might easily happen, I mean as to Mr. Gunter's being then chosen; for that he found him in the chamber of Dr. Brooke (the divinity professor) at the time mentioned, I make no question, by the date of Mr. Briggs's letter, which he had by him, when he wrote his *Apologeticall epistle.* Mr. Gunter afterwards inserted that horizontal instrument in his treatise *Of the sector;* where he acknowledges it was not his own invention, but does not say from whom he had it. When he was settled at Gresham college, his diligence in his profession, and the great improvements he made in mathematical science, soon discovered the right judgment of his electors, and how much they had benefited the public, in their choice of him; for the year following he published his *Canon triangulorum.* This was the first book, that was printed of this kind the nature and use of which the author has himself very modestly described in the following words. *Canon noster usum habet in triangulorum sphaericorum solutione eundem, quen tabulae sinuum et tangentium ab aliis editae; sed praxin paulo faciliorem. Nam eorum multiplicationem per additionem, corum divisionem per subtractionem evitamus. Nec pluribus opus est aut praeceptis, aut exemplis. Idem si desideres in triangulis rectilineis, adjunge nostris amici, et collegae, Henrici Brigii logarithmos. Nam eo nitimur fundamento, eodem utimur operandi modo.* The credit of this improvement of logarithms, in their application to spherical triangles, is given to our author by Edmund Wingate esquire. 'Master Edmund Gunter (sais he) professor of astronomy in Gresham college, in London, hath taken great pains in calculating of a table, containing the logarithms of the sines and tangents of all the degrees and minutes of the quadrant.' The like is done by Mr. Burton in the following passage: 'What so pleasing can there be, if a man be mathematically given, as to calculate or peruse Napier's *Logarithms*, or those *Tables of artificial sines and tangents*, not long since set out by mine old collegiate, good friend, and fellow student of Christ Church in Oxford, Mr. Edmund Gunter, which will perform that by addition and subtraction only, which heretofore Regiomontanus tables did by multiplication and division; or those elaborate conclusions of his *sector*, *quadrant*, and *crosse staffe*.' And the same thing is still more fully expressed by Mr. Henry Bond senior, a noted mathematician in his time, who sais: 'Mr. Edmund Gunter, professor of astronomy in Gresham college, London, first calculated the tables of logarithm sines and tangents to eight places, and caused them to be printed in 1620.' He invented also the *Rule of proportion*, as we are told by Mr. Oughtred who speaking of his own *Circles of proportion*, sais: 'For these, I must freely confess, I have not so good a claim against all men, as for my *Horizontall instrument*. The honour of the invention (of logarithms) next to the lord of Merchiston, and our master Briggs, belonging (if I have not been wrongly informed) to master Gunter, who exposed their numbers upon a streight line. And what doth this new instrument, called the *Circles of proportion*, but only bowe and inflect master Gunter's *line* or *rule*.' In the year 1624 this invention was carried into France by Mr. Wingate, who not only communicated it to most of the chief mathematicians then at Paris, but likewise at their request published an account of its use in the French language; tho this had been then lately done more largely in English by the author himself, in his treatise *Of the cross staff*. But several years after Mr. Gunter's death, Mr. Wingate having translated his French tract into English, published it with many additions and improvements; which has since been called Wingate's *Rule of proportion*, or Gunter's *Line*.

Mr. Gunter likewise drew the lines on the dials in Whitehall garden, and wrote the description and use of them, by the direction of prince Charles, in a small tract, which he afterwards printed by order of his majesty, king James, in the year 1624. The dials were placed, as he describes them, on a stone, which at the base was a square of somewhat more than four feet and an half, the height three feet and three quarters, and unwrought contained above eighty feet, or five tonne of stone. Five dials were described on the uper part; one on each of the four corners, and a fifth in the middle, which was the cheif of all, the great horizontal concave. Besides the dials at the top, there were others on each of the sides, east, west, north and south. But for the several lines drawn upon these dials, and the uses of them, I must refer to the book itself. There was, as he tells us, a stone of the same sise and form, with the like planes and concaves, and dials on them, in that place before; but the lines on his dials were much different, excepting those which shewed the hour of the day. Had Dr. Wallis seen this book, it would have prevented a mistake or two relating to these dials, in a letter writen by him to captain Edmund Halley, dated from Oxford May 23, 1702. 'It was (sais the doctor) about the begining of the reign of king Charles the first, that Mr. Gellibrand (if I have not been misinformed) caused the great concave dial to be erected in the privy garden at Whitehall (which I think is yet remaining) with great care to fix a true meridian line, and with a large magnetick needle, shewing its variation from that meridian from time to time.' The doctor, it seems, had been misinformed, both as to the contriver of that dial, and the time when it was erected; which I thought necessary to observe on account of another important passage in a former letter, writen by him to Dr. Sloane (now Sir Hans) secretary to the royal society, December the 20, 1701, which is this: 'I think it is now agreed on all hands, that what we call *the variation of the variation*, is an English discovery of Mr. Gellibrand (if I mistake not) one of Sir Thomas Gresham's professors in Gresham college, about the year 1635. That is, that the magnetick needle in its horizontal position doth not retain the same declination, or variation, from the true north in the same place at all times; but doth successively vary that declination from time to time. Which tho it were about that time a new discovery, is now admited as an undoubted truth.' In the letter from which the other passage was cited, which as I have said, was writen after this, the doctor himself observes, that the time here mentioned for the discovery of the variation, namely 1635, was misprinted for 1625. If this discovery therefore was owing to one of Sir Thomas Gresham's professors, and made so early as 1625, it must be attributed to Mr. Gunter; and not to Mr. Gellibrand, who did not come from Oxford to Gresham college, till the latter end of the year following, upon Mr. Gunter's death. The stone, on which Mr. Gunter's dials were drawn, remained in its place, at the time Dr. Wallis's letters were writen; but the dials had in a great measure been defaced long before, by the frolics of a certain noble man in the reign of king Charles the second. And the stone it self has now for several years been removed, to make way for the buildings erected in the privy garden, since the unhappy fire at Whitehall, on the 4 of January 1697. There was another very curious set of dials, contrived by Francis Hill, *alias* Line, an English jesuit, and erected in the same garden, on a stone pedestal, in the year 1669. These dials were placed in six ranks one above another in form of a pyramid. But as the surface of them was all of glass, and exposed to the weather, they soon decayed for want of a cover. The contriver published a description of them, a few years after they were set up; at which time they were, as he complains, much damaged. I take notice of this the rather, that they may not be mistaken for Mr. Gunter's dials.

Besides the things already described, he was the author of many other inventions and improvements in the mathematics; most of which were first the subjects of his lectures at Gresham college, and afterwards disposed into treatises, and printed in his works. Had he lived longer, the world would doubtless have reaped more fruits of his fertile invention, and great abilities. But he was taken off the 10 December 1626, about the 45 year of his age, the prime of his time for such studies. He died in Gresham college, and was buried in the church of St. Peter the Poor in Broadstreet, without any monument or inscription; but his memory will always be preserved with esteem by his works, which are these.

1. *Canon triangulorum,* five, *Tabulae sinuum et tangentium artificialium, ad radium* 10000,0000, *et ad scrupula prima quadrantis: Londini* 1620, octavo: 1623, quarto.

2. *The description and use of his majestie's dials in Whitehall garden: London* 1624. quarto.

3. *Of the sector, cross staff, and other instruments: London* 1624. quarto.

All his other works, but that peice of the dials, have passed five editions; the last of which was published by William Leybourn, with additions to several of the books: London 1673, *quarto.* It contains,

1. *The sector, in three books.*

To the third book is subjoined, *The sector altered, and other scales added, with the description and use thereof: Invented and writen by Mr. Samuel Foster.* But more will be said of this in the works of Mr. Foster.

2. *The cross staff, in three books.*

To the second book is added, *An appendix, concerning the description of a cross bow, for the more easy finding the latitude at sea.* And to book the third is subjoined, *An appendix, containing the description of a small portable quadrant: Also, A second appendix, containing the description and use of another quadrant, fited for daily practice: Invented by Mr. Samuel Foster.* Of this also more will be said in his works.

Next follows, *The general use of the canon and table of logarithms.*

3. *Canon triangulorum,* or, *A table of articifical sines and tangents to a radius of* 10,0000000 *parts to each minute of the quadrant.*

To this is added, *Logarithms of absolute numbers from an unite to ten thousand.*

Besides the additional tracts above mentioned, Mr. Leybourn has inserted, as he sais, *divers necessary things and matters through the whole work.* But it is to be wished, he had so printed them, that they might have been distinguished from what belongs to his authors.

IV.

HENRY GELLIBRAND was born in the parish of St. Butolph Aldersgate, in the city of London, on the 27 of November 1597; and in the year 1615 was admited a commoner at Trinity college in Oxford, where about four years after he took the degree of batchelor of arts. He was then, as Mr. Wood sais, esteemed to have no great matter in him; but at length upon hearing one of Sir Henry Savile's mathematical lectures by accident, or rather to save the sconce of a groat, if he had been absent, he was so extremely taken with it, that he immediately fell to the study of that noble science, and conquered it, before he took his master's degree, which was in the year 1623.

While he continued in the pursuit of these studies, the professorship of astronomy in Gresham college becoming vacant by the death of Mr. Gunter, he endeavoured to succeed him. And for that end he procured a testimonial from the president and fellows of Trinity college, which being presented to the electors, he was chosen on the 2 of January 1626. The testimonial was as follows.

'Whereas Mr. Henry Gellibrand hath requested our testimony, the better to make himself known unto such, whose judgements and approbation may further his preferment: We, the president and fellows of Trinity college in Oxon (where these many years he hath made his abode) do testify, that touching his behaviour, he hath been very commendable both in good studies, and virtuous manners; and that more especially his zeale and love for the mathematicks hath been very extraordinary, and indeed very singular among us. In which kind of studies how proficient he is, we referre him unto such as do farre passe us in ability to judge. As for ourselves, we must thankfully acknowledge his very loving readiness, and also dexterity, and facility, freely to communicate to any one among us his knowlege in those studies. And therefore we are full of good assured hope, that for so much as he did long agoe so voluntarily devote himself unto those studies, as that for his affection thereunto he hath contented himselfe with his owne patrimony, and neglected other courses, which commonly others take for their speedy preferment in the world, and for many years hath diligently and entirely applied himself thereunto, and also hath joyned with his private industry much conference with famous professors, both in this university and in London; that upon the examination of the learned he will be found worthy of their good furtherance, and also to have justly deserved our best wishes and endeavours to make him known unto such, as may in any wise pleasure him. Unto whom we do with our affectionate love commend him and them also unto the blessings of the Almighty. Trinity college Oxon, December 22, 1626.

RADULPH KETTELL, *pr.*	ANTONIUS FARRINGDON.
CAROLUS BRAY, *vicepr.*	SAMUEL MARSH.
HANNIBAL POTTER.	ANDREAS READ.
LAWRENTIUS ALCOCK.	GULIELMUS HOBBES.
SAMUEL FISHER.	THOMAS JONES.'

The famous professors of mathematics in London, refered to in this testimonial, and with whom Mr. Gellibrand is said to have conversed, must principally respect those, who at that time were in Gresham college. But Mr. Gellibrand had not been settled there many years, when, as Mr. Prinne informs us, he was brought into trouble in the high commission court by Dr. Laud, then bishop of London, upon the account of an almanack published by William Beale, servant to Mr. Gellibrand, for the year 1631, with the approbation of his master. In this almanack the popish saints, usually put into our kalendar, were omited; and the names of other saints and martyrs, mentioned in the *Book of martyrs*, were placed in their room, as they stand in Mr. Fox's kalendar. This, it seems, gave offence to the bishop, and occasioned the prosecution. But when the cause came to be heard, it appearing that other almanacks of the same kind had formerly been printed, both Mr. Gellibrand and his man were acquited by archibishop Abbot, and the whole court, except bishop Laud; which was afterwards one of the articles against him at his own trial. Mr. Gellibrand was then imployed in finishing the *Trigonometria Britannica* of Mr. Briggs, which was designed by the author to consist of two books. But he dying on the 26 of January 1630, when he had compleated only the first of them, recommended it to the care of his old friend Mr. Gellibrand to draw up the second, and perfect the work. Several other persons also, eminent for their skill in the mathematics, were earnest with him to ingage in this design, which having undertaken and compleated in 1632, it was printed in Holland the following year. He likewise published some other things after this, particularly a discourse *On the variation of the magnetic needle, with the diminution of the variation*, a subject at that time but lately discovered. In this book, for a proof of what he advances, he refers to a collection of *Observations of the variation*, annexed to a treatise of Mr. Edward Wright, intitled *Certain errors in navigation detected and corrected*. Those observations had been made partly by Englishmen and partly by foreigners, in almost all parts of the world, where navigation had then been carried. They have been since much esteemed, and great use has been made of them by very eminent mathematicians.

Mr. GELLIBRAND'S situation at the college, free converse with the lovers of mathematical studies, and diligent enquiries gave him an opportunity of contributing much to the improvement of navigation, which probably would have owed more to him, had he lived longer. But he was taken off more early in life than his predecessor, Mr. Gunter; for he died on the 9 of February 1636, in the fortieth year of his age, and was buried likewise in the church of St. Peter the Poor, without any inscription to his memory. Dr. Hannibal Potter, formerly his tutor at Trinity college, and afterwards president of it, preached his funeral sermon, in which he commended his *piety and worth*. There is a dial made by him, which yet remains on the east side of the old quadrangle in that college. But the best memorial of him are his writings, which are contained in the following catalogue.

1. *Trigonometria Britannica, five, De doctrina triangulorum: Libri duo. Quorum prior continet constructionem canonis sinuum, tangentium, et secantium, una cum logarithmis sinuum et tangentium ad gradus et graduum centesimas, et ad minuta et secunda centesimis respondentia: a clarissimo, doctissimo, integerrimoque viro, domino Henrico Briggio, geometriae in celeberrima academia Oxoniensi professore Saviliano dignissimo, paulo ante inopinatam ipsius e terris emigrationem compositus. Posterior vero usum sive applicationem canonis in resolutione triangulorum tam planorum, quam sphaericorum, e geometricis fundamentis petita, calculo facillimo eximiisque compendiis exhibet: ab* Henrico Gellibrand, *astronomiae in collegio Greshamensi apud Londinenses professore, constructus. Goudae* 1633. folio.

In the year 1658 Mr. John Newton published a folio treatise in English with the same title: *Trigonometria Britannica, or, The doctrine of triangles: In two books. The first* (which was composed by himself) *shewing the construction of the natural and artificial sines, tangents, and secants, and table of logarithms, with their use in the ordinary questions of arithmetic, extraction of roots, in finding the increase and rebate of money, and annuities, at any rate, or time propounded: The second being a translation of Mr. Gellibrand's book* last mentioned.

2. *An Appendix concerning longitude: London* 1633. *quarto*, in three leaves.

This is added to a book intitled, *The strange and dangerous voyage of captain Thomas James, in his intended discovery of the north west passage into the South sea.*

3. *A discourse mathematical on the variation of the magnetic needle: Together with the admirable diminution lately discovered: London* 1635. quarto.

4. *A preface to the* Sciographia *of John Wells esquire: London* 1635. octavo.

5. *An institution trigonometrical, explaining the doctrine of the dimentions of plain and spherical triangles after the most exact and compendious way, by tables of sines, tangents, secants, and logarithms; with the application thereof to questions of astronomy and navigation: London,* octavo.

After the decease of the author this book, having been corrected and inlarged by William Leybourn, was reprinted at London in 1652. *octavo.*

6. *An epitome of navigation.* Also,
Several necessary tables pertaining to navigation. As,
A triangular canon logarithmical, or, *A table of artifical sines and tangents,* etc.
Two chiliads, or, *The logarithms of absolute numbers, from an unite to* 2000.
An appendix, concerning the use of the forestaff, quadrant, and nocturnal, in navigation. London 1674, *etc.* octavo.

Besides these he wrote also some other peices, which have not yet seen the light. As,

1. At the end of his *Trigonometria Britannica* he sais, that he had by him *integram eclipsium doctrinam;* which he designed to have added to that treatise, but that the printer could not wait, till he had revised and fited it for the press.

2. *Astronomia lunaris,* five, *Diatriba in appulsum lunae ad lucidam Pleiadum per triangulorum ratiocinia, e tabulis ac hypothesibus Ptolemaei, Alphonsi, Copernici, Tychonis, Longomontani, et Lansbergii.*

He has himself acquainted us with the time, when this treatise was writen, which begins thus: *Anno* 1634, *Decemb.* 20. *stilo vet. Tubo optico conspexi trientem inferiorem tenebrosae marginis lunaris ingredientem super lucidam Pleiadum, quo tempore deprehendi altitudinem Palilicii* 32 *gr.* 12 *min. Exitum non contigit videre propter nubes debiscentes. Observatio ista habita est Crayae S. Paulini in comitatu Cantii, sub latitudine* 51°. 25′, *et longitudine* 21°, 30′, 5^h. 44′ *a meridie.* He wrote it in about a month, as appears at the conclusion, where he has added, *Hen. Gellibr. Jan.* 22, 1634. And how careful he was to admit of nothing without evidence, he intimates by saying, *Credulitas is mathematico res est summe exitiosa. Rationibus enim verisimilibus errare, quam caeca veritate duci, maluissem.* This book, fairly writen in his own hand, is now in the possession of Sir Hans Sloane baronet.

3. Mr. Wood mentions likewise *A treatise of building of ships*, left by him in manuscript, which after his death came into the hands of Edward Lord Conway.

He had a brother, named John, who lived in Breadstreet, and was his executor. He was the person mentioned by Mr. Prynne, as an evidence at the trial of archbishop Laud.

SOURCE: John Ward. *Lives of the Gresham Professors* (London, 1740), 77–85

41 John Aubrey
'BRIEF LIVES: EDMUND HALLEY AND THOMAS HARIOT'

Edmund Halley (1656–1740)

* Mr. Edmund Hally, astronomer, born October 29, 1656, London – this nativity I had from Mr. Hally himself.

**Mr. Edmund Halley, Artium Magister, the eldest son of [Edmund] Halley, a soape-boyler, a wealthy citizen of the city of London; of the Halleys, of Derbyshire, a good family.

He was born in Shoreditch parish, at a place called Haggerston, the backside of Hogsdon.

At 9 yeares old, his father's apprentice taught him to write, and arithmetique. He went to Paule's schoole to Dr. Gale: while he was there he was very perfect in the Caelestiall Globes inso much that I heard Mr. Moxon (the globe-maker) say that if a star were misplaced in the globe, he would presently find it.

At . . . he studied Geometry, and at 16 could make a dyall, and then, he said, thought himselfe a brave fellow.

At [16] went to Queen's Colledge in Oxon, well versed in Latin, Greeke, and Hebrew: where, at the age of nineteen, he solved this useful probleme in astronomie, never donne before,

☞ viz. 'from 3 distances given from the sun, and angles between, to find the orbe' (mentioned in the Philosophicall Transactions, Aug. or Sept. 1676, No. 115), for which his name will be ever famous.

Anno Domini . . . tooke his degree of Bacc. Art.; Anno Domini . . . tooke his degree of Master of Arts.

Anno . . . left Oxon, and lived at London with his father till [1676]; at which time he gott leave, and a viaticum of his father, to goe to the Island of *Sancta Hellena,* purely upon the account of advancement in Astronomy, to make the globe of the Southerne Hemisphere right, which before was very erroneous, as being donne only after the observations of ignorant seamen. There he stayed . . . moneths. There went over with him (amongst others) a woman . . . yeares old, and her husband . . . old, who had no child in . . . yeares; before he came from the island, she was brought to bed of a child. At his returne, he presented his Planisphere, with a short description, to his majesty who was very well pleased with it; but received nothing but prayse.

I have often heard him say that if his majestie would be but only at the chardge of sending out a ship, he would take the longitude and latitude, right ascensions and declinations of . . . southern fixed starres.

Anno 1678, he added a spectacle-glasse to the shadowe-vane of the lesser arch of the sea-quadrant (or back-staffe); which is of great use, for that spott of light will be manifest when you cannot see any shadowe.

He went to Dantzick to visit Hevelius, Anno 167–.

December 1st, 1680, went to Paris.

* Edmund Haley:– cardinall d'Estrée caressed him and sent him to his brother the admirall with a lettre of recommendation. – He hath contracted an acquaintance and friendship with all the eminentst mathematicians of France and Italie, and holds a correspondence with them.

He returned into England, Januarii 24°, 168½.

Quaere Mr. Partridge of his *Directio mortis,* scilicet about 35 aetatis.

** [Quaere] Edmund Halley who cutts his schemes in wood?
they are well.

[David] Loggan informes me that one . . . Edwards, the
manciple of . . . College Oxon, doth cut in wood very well.

Thomas Hariot (1560–1621)

***** Mr. Thomas Hariot—from Dr. John Pell, March 31,
1680. Dr. Pell knowes not what countreyman he was (but an
Englishman he was) – [There is a place in Kent called Harriot's-
ham, now my lord Wotton's; and in Wostershire in the parish of
Droytwich is a fine seat called Harriots, late the seate of Chiefe
Baron Wyld.]

He thinkes he dyed about the time he (Dr. Pell) went to
Cambridge. He sayes my lord John Vaughan can enforme me,
and haz a copie of his will: which vide.

* Mr. Thomas Hariot – Mr. Elias Ashmole thinkes he was a
Lancashire man: Mr. [John] Flamsted promised me to enquire
of Mr. Townley.

** ☞ I very much desire to find his buriall: he was not
buryed in the Tower chapelle.

*** Mr. Thomas Harriot:– Memorandum:– Sir Robert
Moray (from Francis Stuart) declared at the Royal Society–
'twas when the comet appeared before the Dutch warre– that
Sir Francis had heard Mr. Harriot say that he had seen nine
cometes and had predicted seaven of them, but did not tell
them how. 'Tis very strange: excogitent astronomi.

**** Mr. Hariot went with Sir Walter Ralegh into Virginia,
and haz writt the Description of Virginia, which is printed.

Dr. Pell tells me that he finds amongst his papers (which are
now, 1684, in Dr. Busby's hands), an alphabet that he had
contrived for the American language, like Devills.

He wrote a Description of Virginia, which is since printed in
Mr. Purchas's Pilgrims.

Vide Mr. Glanvill's Moderne Improvement of Usefull
Knowledge, where he makes mention of Mr. Thomas Harriot,
pag. 33.

When [Henry Percy, ninth] earle of Northumberland, and Sir Walter Ralegh were both prisoners in the Tower, they grew acquainted, and Sir Walter Raleigh recommended Mr. Hariot to him, and the earle setled an annuity of two hundred pounds a yeare on him for his life, which he enjoyed. But to Hues* (who wrote *De Usu Globorum*) and to Mr. Warner he gave an annuity but of sixty pounds per annum. These 3 were usually called *the earle of Northumberland's three Magi*. They had a table at the earle's chardge, and the earle himselfe had them to converse with, singly or together.

He was a great acquaintance of Master . . . Ailesbury, to whom Dr. Corbet sent a letter in verse, Dec. 9, 1618, when the great blazing starre appeared,–

'Now for the peace of God[s] and men advise,
(Thou that hast wherwithall to make us wise),
Thine owne rich studies and deepe Harriot's mine,
In which there is no drosse but all refine.'

[Vide] Dr. Corbet's poems.

The bishop of Sarum (Seth Ward) told me that one Mr. Haggar (a countryman of his), a gentleman and good mathematician, was well acquainted with Mr. Thomas Hariot, and was wont to say, that he did not like (or valued not) the old storie of the Creation of the World. He could not beleeve the old position; he would say *ex nihilo nihil fit*. But sayd Mr. Haggar, a *nihilum* killed him at last: for in the top of his nose came a little red speck (exceeding small), which grew bigger and bigger, and at last killed him. I suppose it was that which the chirurgians call a *noli me tangere*.

* Mr. Hariot dyed of an ulcer in his lippe or tongue – vide Dr. Read's Chirurgery, where he mentions him as his patient, in the treatise of ulcers (or cancers).

The Workes of Dr. Alexander Reade, printed, London, 1650; in the treatise of Ulcers, p. 248. 'Cancrous ulcers (*ozana*) also seise on this part. This griefe hastened the end of that famous mathematician Mr. Hariot with whom I was acquainted but

* Robert Hues was buried in Xt. Ch. Oxon.

198

short time before his death; whom at one time, together with Mr. Hughes (who wrote of the globes), Mr. Warner, and Mr. Torporley, the noble earle of Northumberland, the favourer of all good learning and Maecenas of learned men, maintained while he was in the Tower, for their worth and various literature.'

He made a philosophicall theologie, wherin he cast-off the Old Testament, and then the New one would (consequently) have no foundation. He was a Deist. His doctrine he taught to Sir Walter Raleigh, Henry, earle of Northumberland, and some others. The divines of those times look't on his manner of death as a judgement upon him for nullifying the Scripture.

Ex Catalogo librorum impressorum bibl. Bodleianae in Academia Oxoniensi, Oxon., MDCLXXIV:–

Thomas Hariot:– Historia Virginiae, cum iconibus, Lat. per C. C. A. edita per Th. de Bry, *Franc.* 1590 (A. 8. 7. *Art*).

– Same in English, *Lond.* 1588 (E. 1. 25. *Art. Seld.*).

Thomas Hariotus:– Artis analyticae praxis ad aequationes Algebraicas resolvendas, *Lond.* 1631 (F. 2. 12. *Art. Seld.*).

SOURCE: John Aubrey. *Brief Lives,* edited by A. Clark (Oxford, 1898), 282–7.

SELECT BIBLIOGRAPHY

E. J. Dijksterhuis. *The Mechanisation of the World Picture* (Oxford, 1961) provides the best general treatment of science in the seventeenth century and is an indispensable work of reference, as are the *Dictionary of Scientific Biography* and the *Encyclopaedia of Philosophy*. Good introductory studies of the subject can be found in M. Boas. *The Scientific Renaissance 1450–1630* (London, 1962); A. R. Hall. *From Galileo to Newton 1630–1720* (London, 1963); and A. Crombie. *Augustine to Galileo: Medieval and Early Modern Science*, vol II, (1969), which also contains a good bibliography.

The philososphy of science, the nature of scientific method, and the contemporary world pictures, can be studied in their most accessible form in Basil Willey. *The Seventeenth Century Background* (London, 1934; reprinted 1972); E. Tillyard. *The Elizabethan World Picture* (London, 1972); and D. C. Allen. *The Starcrossed Renaissance* (North Carolina, 1941). More specialist and detailed treatment can be found in E. A. Burtt. *The Metaphysical Foundations of Modern Physical Science*, 2nd ed (London, 1932); R. M. Blake, C. J. Ducasse and E. H. Madden. *Theories of Scientific Method: The Renaissance through the Nineteenth Century* (Seattle, Wash, 1960); E. W. Strong. *Procedures and Metaphysics: A Study in the Philosophy of Mathematical-Physical Sciences in the Sixteenth and Seventeenth Centuries* (Berkeley, Calif, 1936); and A. Koyre. *From the Closed World to the Infinite Universe* (Baltimore, Md, 1957), a work that is a most readable classic of the subject.

The changes in astronomy can be traced in T. S. Kuhn. *The Copernican Revolution* (Cambridge, Mass, 1957); Arthur Koestler. *The Sleepwalkers, a History of Man's Changing Vision of the Universe* (London, 1959); S. Drake. *Galileo Studies* (Ann Arbor, Mich, 1970); A. Koyre. *Newtonian Studies* (London, 1965); and I. B. Cohen. *Introduction to Newton's Principia* (Cambridge, 1971).

For medicine, anatomy and physiology G. L. Keynes's *The Life of William Harvey* (Oxford, 1966) and G. Whitteridge. *William Harvey and the Circulation of the Blood* (London, 1971) provide good information, best consolidated by reading W. Pagel. *William Harvey's Biological Ideas* (Basel, 1967). W. Pagel's *Paracelsus* (Basel, 1958) and A. C. Debus's *The English Paracelsians* (London, 1965) are important treatments of their respective subjects, while G. N. Clark's *History of the Royal College of Physicians* (London, 1964) is an indispensable source of reference.

Individual studies that repay attention include F. H. Anderson. *The Philosophy of Francis Bacon* (Chicago, Ill, 1948); M. Espinasse. *Robert Hooke* (London, 1956); R. C. W. Maddison. *The Life of the Honourable Robert Boyle, F.R.S.* (London, 1969); C. Webster. *Samuel Hartlib* (Cambridge, 1970); and F. Manuel. *A Portrait of Isaac Newton* (Cambridge, Mass, 1968).

L. Thorndike. *The History of Magic and Experimental Science* (New York, 1941–58); F. Yates, *The Art of Memory* (1966); and F. Yates. *The Rosicrucian Enlightenment* (London, 1973) provide interesting information upon the less 'rational' elements in seventeenth-century scientific thinking. Paul Hazard's *The European Mind 1680–1715* (London, 1964) is a useful work on the intellectual climate of late seventeenth-century Europe, particularly strong upon the French traditions. R. F. Jones. *Ancients and Moderns* 2nd ed (Washington, DC, 1961) is an indispensable study of English debates in the seventeenth century over the progress of science. J. B. Bury. *The Idea of Progress* (New York, 1960); S. Toulmin and J. Goodfield. *The Architecture of Matter* (London, 1962); and J. D. Bernal. *Science in History*. vol II, 'The Scientific and Industrial Revolutions' (London, 1969) also repay study – the first two for their treatment of their respective themes, and the last as a good outline survey of seventeenth-century science.

Information about scientific academies can be found in M. Ornstein. *The Role of Scientific Societies in the Seventeenth Century* (Chicago, 1913); K. T. Hoppen. *The Common Scientist in the Seventeenth Century: A Study of the Dublin Philosophical Society 1683–1708* (London, 1970); R. Hahn, *The Anatomy of a Scientific Institution: The Paris Academy of Sciences 1666–1803* (Berkeley, Calif, 1971); *Notes and Records of the Royal Society* (1968); and M. Purver. *The Royal Society Concept and Creation* (London, 1967).

There are several good modern reprints of contemporary scientific works. Dover Publications have issued Gilbert's *De Magnete*, Galileo's *Dialogues concerning the Two New Sciences*, Hooke's *Micrographia*, and Newton's *Opticks* and *Principia*. The Johnson Reprint Corporation of New York has reprinted Boyle's *Experiments and Considerations Touching Colours* and Power's *Experimental Philosophy*.

ACKNOWLEDGEMENTS

We are grateful to the following for permission to use material quoted in this volume:

J. M. Dent & Sons Ltd: *The Sceptical Chymist* by Robert Boyle (Everyman's Library, 1967).
Thomas E. Keys: *Classics of Cardiology*, edited by F. A. Willius and T. E. Keys (Dover Publications Inc, 1961).
Penguin Books Ltd: *Descartes' Discourse on Method*, copyright © 1968 F. E. Sutcliffe (translator) (Penguin Classics, 1968).

INDEX

Numbers in italics refer to original extracts